OUR WILD WORLD

FROM THE BIRDS AND THE BEES TO OUR BOGLANDS AND ICE CAPS

ÉANNA NÍ LAMHNA

THE O'BRIEN PRESS
DUBLIN

This edition published 2023
12 Terenure Road East, Rathgar, D06 HD27, Dublin 6, Ireland.
Tel: +353 1 4923333; Fax: +353 1 4922777
E-mail: books@obrien.ie
Website: obrien.ie
First published as trade paperback 2021 by The O'Brien Press Ltd,
Reprinted 2021.
This edition first published 2023.
The O'Brien Press is a member of Publishing Ireland.

ISBN: 978-1-78849-432-8

8 7 6 5 4 3 2 1
27 26 25 24 23

Book cover: art by Linda Fahrlin

Printed and bound by Norhaven Paperback A/S, Denmark.

MIX
Paper from
responsible sources
FSC® C104608
www.fsc.org

Published in

DUBLIN
UNESCO
City of Literature

Enjoying life with
O'BRIEN
obrien.ie

Dedication

For my grandchildren, Archie, Shay and Hugo

CONTENTS

for *Our Wild World*

'Wond   ten both with imagination and feeling ...
a b epr   for living and ... a theory of the world'
Richard Collins *Mooney Goes Wild*, RTÉ Radio 1

'I loved this book ... It should be required reading in schools
and colleges [Éanna] has done a fascinating and excellent job of
encapsulating and explaining very complex situations and processes
and cycles ... a manual to how the world runs'
Niall Hatch, Birdwatch Ireland

'An appealing and insightful take on the natural world ...
a combination of scientific acumen and robust communication skills
are what have made Ní Lamhna one of the most loved educators in
the country. *Our Wild World* lays out some of the basic ecological and
environmental principles that might have passed us by, while clearing
up a raft of myths, everything from migration and bacteria to global
warming and biodiversity. What makes *Our Wild World* particularly
special is that she is that she strikes a tone that speaks to anyone aged
nine to ninety'
Sunday Independent

'[Éanna] talks as fast as she does straight and can convey a lot in
short time ... It's science made simple ... for anyone from about
twelve upwards who has even a passing interest in the twin crises of
biodiversity loss and climate breakdown but is beaten back by jargon.
Experts would also benefit from a read'
Irish Independent

Éanna Ní Lamhna is one of the best-known public figures in Ireland, in particular as a biologist, environmental and wildlife consultant, radio and television presenter, author and educator. Éanna has one of the most instantly recognisable voices on Irish radio and has been for many years a member of the panel of experts on RTÉ's wildlife programme 'Mooney goes Wild'. She also served for five years as president of the national environmental charity An Taisce, and is currently president of the Tree Council of Ireland. Originally from Louth, she now lives in Dublin. Éanna is the author of several popular wildlife books, including *Talking Wild, Wild and Wonderful, Straight Talking Wild, Wild Things at School* and *Wild Dublin: Exploring Nature in the City,* shortlisted for the Reading Association of Ireland Award.

WILDLIFE

We are all in this together

Humans are the cleverest species on Earth – or so we like to think. Although we are less than one million years on a planet that has had life of some sort for the last 3.5 billion years, we have been a most adaptable and successful species, according to our own standards that is. (I wonder what whales or pandas make of us or, indeed, what the dodo or the Tasmanian tiger might have to say if somehow, we could go back in time and interview them.) There are so many humans now that Earth is creaking at the seams. To keep going, we really need to know how the world works. It is not enough anymore to leave this to scientists and specialists.

After all, other species know how the world works and if they get it wrong, they don't last. Survival of the fittest and all that. Hedgehogs that go for a second round of offspring because it has been a really good summer and there was lots of food available, find that the youngsters haven't had enough time to put on the required kilogram of fat to tide them over the winter that inexorably follows. Birds that don't build their nests in safe, hidden spots, well away from the beady eyes of always-watching magpies, may lose the whole clutch in a dawn raid. You've got to know what is going on.

Observing how the world works is fascinating. Nothing beats the evidence of your eyes and finding out explanations for it. But in a world where being fobbed off with a mad explanation on social media can have really significant impacts – as opposed to long ago when fake news meant believing that swans that go missing in summer have turned into beautiful maidens rather than migrated to tundra regions – it is more vital than ever that we all know exactly what is going on and how the natural processes work.

Our planet exists in harmony with the species that inhabit it. While the climate might have been changing over the millions of years, the species on it evolved and adapted to the changing conditions. It was only when a calamitous change happened that species extinction was the outcome. We know that this happened 65 million years ago when the planet was struck by a meteorite and the dinosaurs became extinct as their world had changed quicker than they could adapt to it.

The world is changing very quickly at the moment – according to its own terms of reference as it were. It is heating up exceedingly fast and climatic conditions are affected. This is being caused by one species – humans – who because of their huge numbers and their unsustainable exploitation of the Earth's resources are changing the Earth's atmosphere so that it traps and retains more of the sun's heat. And this is happening in a very short period of time – less than a hundred years. It is calculated that there are less than ten years left at this rate before irreversible change occurs.

We have a beautiful world and we humans are behaving in a beastly way to it. Understanding what is going on is vital for everyone so that the steps needed to sort this out can be taken and supported. It doesn't have to be a technical, scientific explanation – a humorous simple scientific explanation does the job too. We share our world – there are other ways of living. We do need to understand this. We are only one species and yet so often the attitude to an unrecognised fellow species and sharer of our world is 'what is this and what will it do to me?' or 'what is this and how do I get rid of it?' It's not all about me actually – it's all about us all. We are all in this together.

THE GREAT OUTDOORS

How the world works

If we don't know how the world works, how can we possibly know how to behave in it? By a series of amazing co-incidences, planet Earth is suitable for living things to exist on it. This is so remarkable that, as of yet, in all the galaxies in space that we have discovered, we have not found a single other place where we have detected living things. There are no other planets to which we can migrate when we have banjaxed this one. The mad money being spent travelling to outer space would be much better spent keeping this planet habitable for life here. Imagine the terror among living things on some other planet, if such an inhabited one is ever found by us, when they see us coming and learn what we did to our own beautiful Blue Planet.

Viewed from space, it is indeed a blue planet because two-thirds of its surface is covered with water. We are the only planet that we know, so far, that has water and the conditions suitable for retaining it. Water plays a vital role in maintaining life on Earth.

The way the world works is beautiful in its simplicity. It is heated by our star – the sun – from which we are just the right distance to get the right amount of warmth. Too near and, like Mercury and Venus, it would be far too hot. Too far away, like Mars, Jupiter and Saturn, and it would be ever-lastingly cold. But like in the fairy story about the greedy Goldilocks who wanted to gobble the porridge she found in the house of the three bears, the temperature of Baby Bear's porridge (and our planet Earth) was just right. And believe it or not, this co-incidence that finds our planet getting just the right amount of heat to support life is called the Goldilocks effect.

Our atmosphere is just right too. First of all, we actually have one. It consists of gases in the right amounts – nitrogen, oxygen, carbon dioxide, water vapour and a few rare inert gases. The percentages of these have changed, albeit slowly, over the 4.5 billion years of the Earth's existence. There is an equilibrium between life on Earth and the atmosphere around it. Or there was, that is, until one species – humans – got too big for its boots.

We have several broad categories of living things – plants, animals and microbial life, such as bacteria. The interaction between all these is what keeps the whole show on the road. The sun provides all the energy needed. Plants make all the food for the other two groups, using the energy in sunlight to fix carbon from the carbon dioxide in the air to make carbohydrates and from them other more complicated food groups. A by-product of this is oxygen, which the other two groups need to survive.

Animals depend on the plants for their food. Some eat the plants directly and rejoice in the swanky title of herbivores. Others kill and eat these herbivorous animals – these are the carnivores. And some very well-adapted animals can digest both the plants and the meat from animals and go by the accolade omnivores.

The third group – bacteria *et al* – are the decomposers. They break down dead plants and animals and release the carbon back into the atmosphere as carbon dioxide – ready for the plants to take it in as they grow. A very elegant and stable cycle indeed.

There are a few rules as it were. All the food is made by the plants and the energy it provides passes up through the food chain, diminishing as it goes. Herbivores don't eat ALL the plants in a balanced world or they would quickly run out of food as there would be no seeds from which new plants could come. Carnivores don't eat all the herbivores either. The rules of ecology say that the predator is controlled by its prey, not the other way round. The predator picks off the weak, the slow of foot, the unwary, but the smartest and best adapted survive – the survival of the fittest.

And no carnivore will expend more energy catching its prey than it gets from the kill when it finally makes it. While cheetahs can run faster than wildebeests, such fast running uses up a huge amount of energy. It cannot afford to chase the poor beast all day. It needs to sneak up carefully and then, with a final burst of speed, nab the unwary animal. If it misses, it doesn't pursue the chase; it shrugs its shoulders, says 'ah well' (or something) and waits for another to come along. Hounds are only able to pursue a fox for a long time in the so-called sport of foxhunting if they are first well fed in the kennels. If the dead fox were their only sustenance, they couldn't do it.

In good times, when it all works well, animals thrive, come into heat and reproduce. But there can't be enough food for all the offspring to survive or else the numbers would increase exponentially. Imagine the two robins in your garden building their nest and being watched so fondly by yourself. Four eggs, four chicks, and in a half-good year, robins will nest twice. (In a really good year they might even nest three times, but let's just go with twice.) So at the end of the summer, if the rules of ecology were suspended, you would have ten robins in your garden – Daddy and Mammy and eight babbies – five times the original number. And going with this scenario there would be ten robins in each robin territory in the parish. Robins can live for ten years. If they reproduce at the same rate, there will be 50 robins in your garden in a year's time, and 250 the following year. By year ten the number is just short of 4 MILLION. But, actually, in year ten there are only two robins in your garden. And there is not a heap of almost 4 million dead robins piled up on the front lawn either. The truth is that for the population of robins to keep steady there must only be two robins the following spring to breed in your garden. One of the adults and one of the eight offspring will have survived, and no, they are not here in your garden breeding with each other, they have found new partners over the winter. Only the smartest youngster has survived. The others have become food for carnivores higher up the food chain or they have died of hunger, being unable to find food in the winter, as also happened to one of the original two adults. I am not making this up – bird ringing studies over the decades have provided chapter and verse.

And it is the same for all the species. A butterfly will lay 100 eggs on your prized cabbages. If one or two of them succeed in becoming a butterfly, the species is stable. The other 98 caterpillars will have gone to feed the baby robins. You can see how a food chain easily becomes a food web, as baby robins are fed other things too, such as spiders and small worms. And that is how it works. The availability of food controls the number of individuals that can survive. Animals of course have developed techniques over the ages to make sure to survive. Move away if all the food here is gone. Sleep if there is nothing to eat, and drop your metabolic rate so that you need less energy when times are very hard. Eat something else too; don't just depend on one food. If you can't do any of these, then indeed you might become *aimsir caite*, as happened to the giant Irish Deer. On the other hand, sharks are in such perfect equilibrium with their environment that some species haven't altered at all in millions of years.

Humans began to evolve about three quarters of a million years ago. They are omnivores, and for most of the time since then, they were hunter-gatherers. Palaeolithic (Stone Age) people and Mesolithic people hunted animals and gathered plants, living in relative harmony with the earth, as indeed some forest tribes in tropical rainforests do to this day. Until ten thousand years ago we were all hunter-gatherers, and our whole world population was estimated to be around 10 million, give or take a few million.

And then we discovered farming. We could control the supply of our food. Someone – probably a woman – noticed that some of the grains that she had been painstakingly collecting and grinding into flour of sorts had not been ground up but abandoned somewhere and grew into fat little grasses with more of the same seeds. Get enough of these, plant them yourself and wait a season and then there wouldn't be all this traipsing around gathering and moving on when all was collected, Why, she could build a nice permanent tent and send the children to school. Similarly, the young of the herds that the men spent so long away hunting, could be kept around the place and domesticated. And the men would then be there to help with everything. Yes, in theory, it was a great idea.

Farming began in the Fertile Crescent around the Euphrates River in the Middle East about ten thousand years ago and humans felt that they were sorted. All you needed was land to farm. They increased and multiplied because they could grow enough food for all, and they began to spread out seriously. As the last Ice Age retreated, lands in Europe became available. The first farmers – Neolithic people – to arrive in Ireland made it here 5,000 years ago and we have evidence of their first fields and what they farmed, in the Céide Fields of north Mayo.

Meanwhile, the rest of the species on Earth behaved as they always did. Which is just as well, as life on Earth is only possible with the co-operation of all the species here. Let's look in more detail at what is involved and the interesting ways our fellow species live.

2

POLLINATION

Another name for the birds and the bees actually

I have a friend who a few years ago got himself a polytunnel. He was delighted with it. At last, he could grow the things that it was always too cold or too windy to grow outdoors. Strawberries were high on his list. His outdoor attempts before this had always been gobbled by slugs before they even ripened, or else the blackbirds finished off the rest. Now with his polytunnel, such marauders would be kept at bay. And, indeed, his strawberry plants grew very well and had lots of flowers in early May. Things were looking good. But he didn't get one single strawberry fruit on any of the plants. I was duly called in to advise. His polytunnel was nice and hot; he had watered everything with due diligence; it was in a bright spot in the garden so there was plenty of light. Did he leave the two ends open each day, I asked him? Of course not, he didn't want birds flying through gobbling his strawberries or slugs slithering in either. So, nothing flew in at all. How did he think his strawberry flowers were going to be pollinated if he kept all the flying insects out? No insects, no pollination, no strawberries. Time for a chat about the facts of life.

Why do plants have flowers anyway? The answer is that this is where their reproductive parts are. Many flowers have both male and female parts in the one flower. The male parts are the stamens and at the top of each stamen the pollen is stored. The female part is in the centre of the flower and is called the ovary. A long tube called the style – there can be more than one – leads down to the ovary. And, indeed, some flowers may have more than one ovary. But the basic fundamental principle is the same whatever the individual arrangement is. The male pollen must go into the female ovary and fertilise it before seeds can form. Naturally, in the interest of genetic variation, flowers don't pollinate them-selves. (There are of course exceptions, but in general this is the case.) So, to avoid mischance, the pollen and ovary in any particular flower are not ripe at the same time. Pollen must come from another flower and land on the receptive stigma at the top of the style in order to fertilise the flower.

But pollen cannot move by itself. Something has to bring it. And at our latitudes this is done by insects – pollinating insects or pollinators. In tropical countries some pollination is carried out by birds, such as hummingbirds, or by fruit bats, but here we are dependent on our pollinating insects. Why should our insects pollinate our flowers? What is in it for them? Like any good trade union member, they are not going to work without pay. But actually, they don't have a meeting every morning and say we must go out and do our good deed for the day and pollinate plants. If they have meetings at all, they decide to go out to the pub for a pint. Because that is why most insects visit flowers – for the free drink that is nectar.

Flowers have evolved together with insects and they depend on each other. The lovely petals and the smells produced by the flowers are to attract the pollinating insects. The pay for the job is a drink of lovely sweet nectar. But flowers are no fools. They keep their nectar well hidden, deep down in the flower, and the visiting insect has to stick its head right in, in order to suck up the lovely liquid. But in truth it is not much of a pub. There is only one small drink on offer and so the still-thirsty insect has to go on a pub crawl, as it were, of many flowers to get enough nectar.

As it goes down deep into the flower, it rubs its hairy head off the ripe pollen which stays on it as it flies off to the next flower. Here, it might be the female ovary that is ripe, and the nectar-seeking insect accidentally deposits the pollen from the last flower on the sticky stigma as it quests for more nice nectar. And BINGO, fertilisation happens.

The insect doesn't care; all it wants is the nectar. The head covered in pollen is an awkward side effect as far as it is concerned. And this is the story with all the butterflies, hoverflies, wasps, and other flies that visit flowers. They are only after the one thing … nectar.

The only group of insects that actually wants pollen is the bees. It is a nuisance to everyone else. But bees collect pollen because it is rich in protein and they bring it back home to feed the baby bees in their larval state. The bees have modified hairs on their back legs – pollen baskets – which they load up with pollen collected from the flowers they visit. They too have to visit lots of flowers to collect a full load, and as they do, they carry some from one flower to another and fertilisation takes place. A loaded honeybee can carry up to a third of its weight in pollen back to the hive – the equivalent of a human with 30kg of luggage. They spend three weeks of their life collecting pollen to feed the young and then they switch tasks to collecting nectar to bring home to make honey – the food of the adult bees. Again, this requires the head to be inserted deep into the flower and the nectar sucked up into the bee's nectar sac. To fill the sac, several flowers have to be visited and of course pollen adhering to the hairy head of the bee will be transported and deposited as she does this. In this way bees do twice as much pollinating as the other groups of insects. No wonder she drops dead of exhaustion after six weeks' work – three on pollen-collecting duty and three on nectar collection. Being as busy as a bee is not necessarily a good way to be described if you are a human!

My friend's polytunnel was never visited by any flying insect and so no pollination of his strawberry flowers ever took place and therefore he had no strawberries. (But the carrots and the lettuce and scallions did well, so I didn't leave empty-handed.)

Some very sophisticated flowers have evolved to be pollinated by the birds, not by the bees. Well, not any old bird species, but by birds that depend on them for nectar, as they have no other source of food. The hummingbirds of North America are a wonderful example of this. Why are they called hummingbirds? Have they forgotten the words of the song and have to hum the tune instead? Well no, actually. The humming sound comes from the very fast movement of their wings – up to fifty beats per second – as they hover in front of an attractive flower.

An attractive flower to a hummingbird is one that contains lovely nectar, and so it inserts its long thin beak into the flower and drinks the delicious liquid. Not enough, alas, in one flower to satisfy a hungry hummingbird. So, it flies to the next flower for more, not knowing or caring that its head is covered in pollen from the previous flower. And as it greedily guzzles more lovely nectar from deep within the next flower, it deposits the pollen on the female stigma and fertilises the second flower.

Sunbirds do the same services for flowers in Africa and southeast Asia. They don't hover in front of flowers as the American hummingbirds do. They perch on the flower stem and drink the nectar at their leisure. But they still end up with the all-important pollen on their heads and complete their side of the bargain.

Mind you, some hummingbirds can be too smart for their own good. The robber hummingbird, which lives, among other places, in Costa Rica, has worked out that it can get at the nectar by going around the back of the flower. It flies to the bottom of the petals and pokes a hole with its beak directly through the petal bases into the nectar well. It can drink all the nectar directly through this back door, as it were, without getting its face and feathers all covered in that dusty pollen. These hummingbirds get all their nectar this way. They have concentrated their entire efforts on one particular species of flower that is easily poked open through the petals at the back. Clever or what?

In fact, it is certainly not clever. What these humming-birds don't realise is that flowers reproduce by sending their pollen off to the female parts of other flowers using the hummingbird's facial feathers as the means of transport. As far as the flowers are concerned, they are paying for these sexual services by providing a meal. However, because the smarty-pants hummingbird has short circuited the process the flower's bribe, the nectar, is being taken but the pollen delivery price is not being paid. Before long, these flowers will become extinct as new seeds are not being made. And there will be no more wells of nectar for these humming-birds to feed on as they are specific feeders on these flowers. Both are on the road to extinction. A prime example of how you can be too smart!

What happens if a plant finds itself in another country where its pollinator doesn't live? Well, one of two things as you might imagine. It doesn't survive as pollination doesn't happen, or it succeeds in persuading a savvy local to do the job. This is what the New Zealand flax has done. This big, strong, hefty plant is a close relative of the day lily, not of flax. It was introduced originally to these shores as a potential fibre crop, but it was soon discovered that the tough leaves were too difficult to break down to release the fibre, and cheaper sources of fibre were found.

With its robust, stringy fibres, the New Zealand flax is commonly used as a hardy garden plant, and in Ireland's western counties near the coast it is chosen as a windbreak around fields. There, it succeeds in surviving the strongest gales and even the salty sea-spray and wind-blown sand.

In its native home of New Zealand, this plant is mainly pollinated by a native bird, called a Tui, that feeds on nectar and which has evolved a perfectly shaped beak to fit with the curvature of the flax flower. As we don't have Tuis here, or indeed any native flowers requiring birds for pollination, there were no worries that the New Zealand flax would become an invasive species, spreading in the wild.

But of course, Murphy's Law applies. It wasn't long before sharp-eyed botanists began seeing it grow where it had never been planted by gardeners. How had it self-seeded? Then new birds were noticed by the birdwatchers – birds not unlike starlings, but with bright orange foreheads. And yes – you've guessed it. Smart hungry starlings had learned to perch on the robust stems of the flax and bury their heads into the flowers to drink the lashings of lovely nectar inside. No such thing as a free meal and they were liberally daubed with orange pollen by the flower as they did so. Evolution happening before our very eyes.

Birds do it, bees do it, and in some parts of the world, bats do it. Much of the banana supplies of Europe come from countries in Central America – Costa Rica and Nicaragua. If you visit the banana plantations there, you will notice that the ripening bananas on the plants are often covered with blue plastic covers. Why, can't the growers wait till they harvest them before wrapping them up? Bananas, to our eyes, grow upside-down, as it were. The large flower is beneath the bananas and the fruit ripen from the top down, so the ripening bananas are on the top and the bananas are pointing upwards. The banana flowers open at night, and their lovely sweet nectar is eagerly sought after by the night-flying fruit bats. They hang upside-down from the ripening bananas to get right in at the lovely nectar. And they leave little footprints all over the good bananas. We are such fastidious customers in Europe that we expect our bananas to be perfect, with no bat footprints. The growers need the bats to pollinate, so they have come up with the ingenious solution of protecting the fruit by covering their bananas on the stalk.

Some flowers don't bother with all this palaver at all but depend on the wind for pollination. There is no need for attention-grabbing petals or seductive perfume. No need to get a drinks licence either as there is no nectar. Many wind-pollinated flowers are either male or female. The male ones can be in the form of long catkins, full of pollen that shake in the wind and release clouds of pollen to, hopefully nearby, receptive female flowers. Trees such as hazel and willow have wind-pollinated catkins. All conifers are wind pollinated too. And, indeed, so are all members of the grass family, although their flowers all have both male and female parts. The abundance of pollen released by fields of long grass in early summer can cause hay fever to those of us allergic to pollen.

Much of our food comes from plants that are pollinated by pollinators. All our fruit and our above-ground vegetables. Even vegetables that come from other parts of the plant, such as bulbs, tubers and root vegetables, come from plants that are insect pollinated to provide the seed in the first place. Crops such as soya, rapeseed, and almonds depend on pollinators. In fact, remove all these and the world's supply of plant food coming from wind-pollinated plants would just be cereals and some tree-borne nuts. Rice, maize and wheat are vital crops to feed the human race, but not on bread alone doth man live.

Honeybees and the wild pollinators, such as bumble bees, solitary bees, butterflies etc, need suitable environmental conditions to thrive. They all hibernate in some form, except the honeybee, which is awake but indoors in its hive during winter. So, when they get going again in spring, they need lots of food to become established and start breeding. They need flowers with pollen if they are bees, and nectar-rich ones are needed for everyone else too. Oh, and they have to be able to see them as well. Bees cannot see the colour red, so while humans might like great expanses of red flowers, they are no good for the bees who cannot see such a colour, although they can see in ultra violet at the other end of the spectrum, which we can't.

The wildflowers that evolved in harmony with the pollinators are not red. Even the wild poppy is an orangey colour and has UV lines guiding the bees in to the centre, where the pollen and female styles are. Indeed, some of our most cherished spring garden flowers, such as daffodils and tulips, are no good at all to pollinators as they have no nectar and so provide no food for them.

Modern farming methods have an impact on our pollinating insects too and usually not in a good way. Grasslands for grazing cattle are just that – fields of grass with as little ground as possible taken up by other plants, such as meadow flowers. Fields are bigger too than they used to be, with fewer hedgerows and their flowering shrubs and flowers. A particularly nasty type of insecticide has been invented – neonicotinoids. These work by making the whole plant poisonous to insects, including the pollen, so visiting bees end up collecting poisonous pollen. We have banned these in Europe but not in other parts of the world. So, it is not surprising that intensively grown almond orchards in

California, which have all the 'weeds' under them sprayed and removed, are putting huge pressure on the very insects that are needed to make the nuts form in the first place.

It is great to see increased awareness of this situation here in Ireland. Our All-Ireland pollinator plan has drawn attention to the plight of the bumble bee and its colleagues. Flower-free lawns and tightly mown grasslands in parks are beginning to give way to wildflower meadows that are much more favourable to pollinating insects. Hanging baskets on town and village streets are less likely to be full of red begonias and petunias and more inclined to be sporting purple and yellow flowers that contain pollen and nectar. We are becoming more aware that we not only have one species of honey bee and twenty-one species of bumble bee , but that we also have about seventy-eight species of solitary bee as well – the single mothers as it were of the bee world. While these don't live in a colony with a queen, they still provide their offspring with pollen supplies to eat when they hatch out in their individual nests, and they also need nectar to keep themselves going as adults too. Wildflowers for all these as well too please.

Albert Einstein is often quoted as saying that if bees become extinct then so will humans too, within four years. Mind you, I don't know of any research he published to back up this claim or, indeed, if he ever made the claim. After all, his specialist field was relativity not botany. But even if all the pollinators were gone, we would still be grand. We could all manually pollinate our crops with feathers, gently moving the pollen from one flower to another or just live on bread and rice and corn. Couldn't we? Couldn't we?

3

MIGRATION

**Why do they go and, more importantly,
why do they come back?**

Migration — the great global movement from north to
south and back again — happens every year and lifts our
hearts. The first swallow is the sign that summer is on the
way. The first call of the cuckoo has inspired poets from
one of the greatest of the wandering Irish bards Séamus
Dall Mac Cuarta, the blind 17th-century south Ulster
poet, to William Wordsworth. Mac Cuarta's poem, written
around 1707, begins '*Fáilte don éan is binne ar chraoibh*' —
Welcome to the sweetest-sounding bird on the branch,

while Wordsworth wonders 'Oh cuckoo! Shall I call thee Bird or but a wandering Voice?' The return of the stork every year to the large chimney nests in continental Europe – bearing newborn babies to happy families – is the stuff of legends. But not all birds migrate, only some do. Why? Where do they go and, even more importantly, why do they come back?

In Ireland, we have several different bird migrations. The most familiar one is the return of birds from warmer climes to breed here in summer. The swallow is the one everyone knows, but similarly designed birds include the house martin, the sand martin and the swift. These are all magnificent fliers and aerial feeders. They fly around like little hoovers with wide open gapes, collecting the insects in the sky as food. They share the aerial space between them, swallows lowest, house martins above that and the swifts in the higher section of all, while the sand martins patrol the skies nearest to their nests, which they make in tunnels on exposed sandbank faces.

The swallows are the first to arrive. They always return to the parish where they were born – as ringing by bird experts has proved. They nest indoors in farmers' sheds and barns, which has given rise to their name – barn swallows – in other countries which boast of more than one swallow species. They return to the same barn, and the same nest, which they repair with mud, collected from wet muddy patches around. They can raise two or even three broods over the summer, feeding them with balls of flies, which they accumulate in their aerial sorties.

The other one – the house martin – is familiar to us as well. In fact, lots of people don't realise that there are two different species – the swallow, which always nests indoors, and the house martin, which sticks its mud nest to the wall of the house, beneath the apex or just under the eaves above the door. This is not a favourite site for the householders who can be festooned with droppings from the birds above if they emerge from the house at the wrong time. These house martins also return to the same sites year after year and often build nests beside each other – colonial living. To identify the house martin look for the shorter forked tail and the very obvious white back just above the tail, easy to spot as the bird flies.

Our third visitor in this category is the swift – the last to come and the first to leave. Swifts are always with us by 15 May and are gone by 15 August – well in Dublin anyway round where I live. On a bright May evening when you are outdoors enjoying the sunshine, particularly in towns and cities, you suddenly hear the shriek that you haven't heard for almost a year and look up and there it is. The scimitar-winged, fast-flying swift, quartering the sky above.

Final proof that summer is here. Swifts are absolute masters of flight. So much so that they are completely at home in the air. They get all their food there. They even sleep there. Not like us with our requirement for eight hours of sleep per night – no, if they did that they would fall down. They sleep in a different way, high up on the thermals that keep them aloft, for seconds at a time. One half of the brain sleeps while the other half remains awake.

They can even mate on the wing too. In fact, the only thing they can't do in the air is lay an egg. Well, they probably could, but that wouldn't help the survival of swift populations. They need to lay their egg in a secure place and sit on it to hatch it out, like any sensible bird. They are four years of age before they do this, and as they have never been down from the skies their leg muscles are weak, and they cannot walk. Were they to land on the ground, it would be curtains for them; they couldn't ever take off again. So, they seek space in eaves and soffits where they can swoop in from flight, land and lay their egg. Then they shuffle to the edge again and launch themselves into the air from that height. So, no nest building. How could they gather material? Maybe a few of their own feathers, but that's it. It takes about twenty days to incubate the eggs and then five more weeks feeding the young with flies collected from the air before it can leave the nest and fly too.

We have other summer visitors as well – cuckoos, corn-crakes, wheatears, flycatchers, chiffchaffs – all of which come here to breed. What they all have in common is a dependence on the great supply of insects that are available to feed the young that they produce during their breeding summer. Once the days begin to shorten they can detect the lessening daylight hours, and they fatten up and fly off to where the supplies of insect food are again guaranteed.

Where do they go and how do they find the way? In the main, the young birds fly with their parents, who have made the journey already and know the way. But where do they go? This was always a mystery to people long ago. It was said that swallows dived down to the bottom of ponds and remained there in suspended animation until early summer arrived again and they rose from the watery depths. But, thanks to the efforts of our bird ringers, we now know that this is not so. Swallows make the long journey across the equator to South Africa, where it is summer during our winter, and they can continue their aerial feeding in the summer skies there. They have learned the route from the generations of swallows before them, whom they accompany on their first journey southward and then they, in turn, remember how to navigate it.

Which is all fine until you consider the case of the cuckoo. As we all know, the cuckoo lays her eggs in other birds' nests

— an action that sees them branded as brood parasites. Our cuckoos in Ireland seek out the nests of the meadow pipit, in which to lay. As the female can lay up to nine eggs and can only leave one in each nest, she must find nine different nests to lay in. Having been raised by meadow pipits herself, she knows exactly how to find them. In Britain reed warblers and dunnocks are parasitised too, with particular females specialising in one species of host. This all happens very sneakily throughout May and June. Then, job done, there is no longer any need to hang around. The raising of the young will be carried out by the poor overworked foster parents, so the adult cuckoos can return to their wintering grounds, apparently in the jungles of the Congo, as early as July. The young are successfully reared by the foster parents, who being much, much smaller that their grotesquely sized squatter, must be worn to a flitter finding enough insect larvae to feed the usurper. The young cuckoo never sees its biological parents, never hears its father's two-note song. How does it even know it is a cuckoo, not a meadow pipit? It must do, because every year in September when it is fledged and fully reared, instead of moving to the next field or the next parish as the young meadow pipits do, this young cuckoo heads off for Africa. Why? What makes it do it? How does it even know that Africa is there? How does it know to stay in the tropical jungles of the Congo River — a far cry from the upland grasslands or the Burren limestones where it was reared? It is all hard-wired into its brain, part of its genetic inheritance. Amazing.

We have migratory birds too that come here just for the winter. These are birds that breed in the high Arctic tundra and feed on the vegetation – the grasses and other plants that grow there. Geese, such as Barnacle, and White-fronted geese that breed in Greenland, Brent that nest in Arctic Canada and Whooper swans from Iceland all cross the Atlantic to come here in October as their feeding grounds become covered in snow and the vegetation is unavailable to them. We have visitors from further north and east in Europe too. Wading birds with long legs and bills, such as curlew, dunlin and godwit, who feed by probing their bills into the soft mud, cannot do so when the mud is frozen rock hard and so flock to our muddy estuaries in winter, where an abundant supply of food awaits them.

Again, these winter migratory patterns were noticed by people long ago and quite often they had no rational explanation for it. Our Mute swans – the breeding ones with the orange bills – are relatively recent arrivals in Ireland. They are thought to have been brought in by English landlords as ornamental species for the ponds and lakes on their estates. In England they were domesticated in the twelfth century and were the property of the Crown until the eighteenth century.

The swans of our myths and legends were the Whooper swans. The Children of Lir in the legend of the same name were changed into swans by their wicked stepmother, but retained the ability to sing. Whooper swans sing, so it figures that the children of Lir could sing as swans. And their comings and goings reflect the fact that Whooper swans are migratory birds, vanishing for the summer and returning with the longer autumn days.

The old Irish did not know where swans went in the summer. As far as they knew, there was no land over the ocean to the west and so this gave rise to many stories where swans turned into beautiful maidens and indeed vice versa. It was thought that they embodied the souls of virtuous maidens and that they could turn from swans into women. There are various stories about men finding beautiful women bathing in lakes. All goes well, indeed very well, children may even be born, but the one thing the husband must take great care about is never to let her have her original cloak, the one she wasn't actually wearing when he found her bathing. Why the men in these legends never destroy the thing in the fire is beyond me because invariably at some stage the woman finds the cloak, puts it on, turns into a swan and flies away.

Lots of stories centred on Tara involve swans. King Eochaid wagered a hug (or something) with his wife in a game of chess against Midir and lost. When Midir came to Tara to claim his hug there was great security around the whole place. Nothing daunted, Midir clasped Étaín to him and they both rose through the smoke hole in the ceiling and flew away as swans. It was nine years before Eochaid got Étaín back. Étaín's grandson Conor, unaware of the family story – or maybe only too aware of it – hunted swans, until one fine day a flock he was chasing turned into armed men who told him that it was very wrong to kill birds. They also told him if he walked naked to Tara, carrying only a sling and a stone, he could become king. Which he did. Which just goes to show – don't mess with flocks of Whooper swans.

And do swans sing just before death – the swan song? Of course not, but the story goes back to the Roman Pliny the Elder. The swans sing, according to him, not in grief but for joy that they are at last going to meet the god Apollo, whose birds they are.

Hans Christian Anderson has stories about swans too – the Wild Swans which move from one country to another, while the Grimm brothers had a somewhat similar tale called the Six Swans.

In days gone by, the absence of these large birds for half of every year had to accounted for in some way and as the only difference between science and magic is that there is no explanation for magic, then these legends and stories provided a satisfactory explanation.

This, then, is generally the pattern with migration. Birds move south to feed when their food is not available further north. In winter when there are no insects for swallows in Ireland, no grass for Whooper swans in Iceland, and hard frozen mud that no wading bird's beak can access in Siberia, then it is all perfectly understandable – fly south or starve.

The real question is why do they make the return trip? Why do Barnacle geese undertake the hazardous journey back across the ocean in April? We have perfectly good grass here all summer. Our country is famous for it; 55% of our land is covered in grass. What makes the swallow undertake the arduous journey through Africa, over the Sahara and much of Europe to reach a shivering Ireland, lashed by the cold easterly winds of March? There are plenty of insects in Africa – no need to come all this way, you might think.

They do it for the lengthening day. Birds (except for owls and a few other nocturnal species) are daytime feeders. They fly north to breed, and hungry chicks need a lot of nourishment to grow. So, swallows need the eighteen hours of daylight we get here in high summer to catch enough flies to feed their fast-growing offspring. There is not much nourishment in grass; you need to eat lots and lots of it if it is your chief source of nourishment. (With their rapid metabolism, geese process large quantities of grazing, doing a dropping every seven minutes.) So, the young goslings need the twenty-four hours of daylight that summer north of the Arctic Circle provides in order to grow to maturity and be strong enough to make the journey south when conditions deteriorate.

Climate change does not affect the movement of the Earth around the sun. The equinoxes are always in March and September and the solstices in June and December. Birds' brains have evolved to be stimulated by the change in the length of daylight and it is this that triggers these great migration events. But if temperature changes are beginning to mean that food is no longer available when the weary visitors arrive, can they evolve quickly enough to adapt to the changing conditions?

There are other migrations too that have different triggers. We have a summer migration all along our rocky coasts of seabirds, many of them in densely packed, noisy, smelly colonies. Seabirds such as gannets are supremely adapted to living on the sea. Gannets have magnificent eyesight and can spot shoals of fish from great heights and dive down upon them like arrows with their wings folded for aerodynamic smoothness. In fact, they have reinforced skulls, so that they don't brain themselves when they hit the water. Their original Latin name *Sula* comes from the Irish word *súil*, meaning an eye, reflecting how keen-sighted these birds are.

They can live all their lives perfectly well at sea, but the one thing they cannot do there is lay eggs and breed. All sea-birds must come ashore to do this. Hence the great summer breeding migrations. Thousands of pairs of gannets crowd the rocks of Little Skellig, off the south coast of Ireland, every year to breed, the second largest breeding colony of gannets in the world after St Kilda in Scotland. There is safety in numbers, so woe betide any marauding gulls that think they can snaffle a few eggs. But many seabirds that are highly adapted to living at sea find it very difficult to manage on land.

Puffins have to resort to flying in on short stubby wings that are more adapted to swimming underwater than flying. They nest in commandeered rabbit burrows – all tactics to keep them safe from the ever-hungry, aggressive gulls.

Storm petrels and shearwaters are so uncomfortable on land that they only visit their hungry offspring under the cover of darkness to feed them.

Fulmars have their legs positioned so far back on their bodies that they find it difficult to take off if they have landed. They nest on small ledges high up on cliffs and launch themselves into the updrafts in order to fly away again.

Brave and hungry island men in times past used to descend these breeding cliffs on ropes in high summer and collect eggs and fat young birds for food to tide them over hard times. Puffins are still eaten by some in Iceland, where they are known as *lundi*. Wild populations of anything have to be harvested in a sustainable fashion if they are not to be endangered. Traditional hunters knew this instinctively. Iceland puffins are not being endangered by the amount of hunting that is carried out for puffin dinners. Modern fishing trawlers with sonar to detect fish shoals are not playing the game, however. If the seas are being stripped of the fish stocks needed to support these seabird colonies then their numbers will inexorably decline.

Mammals can migrate too. Great herds of wildebeest migrate in east Africa. Their seasons are determined by the rains that cause the growth of grass on the savannahs where they live. Caribou herds in Arctic Canada, which also feed on vegetation, move south to avoid the cold, covering snows of winter and north again when the melt comes. It's one way of coping with the cold of winter. However it's not the only way.

Other creatures have different strategies – hibernation for one.

4

HIBERNATION

The ultimate cop-out

Surviving winter is difficult for wildlife. Food is scarce and more energy is needed to keep warm. Those that can follow the warmer weather. It's easy for birds. They can fly off to hotter climes. But what about mammals whose food supplies vanish when the colder weather comes? What are they supposed to do? A wonderful strategy is to sleep it out for the winter and wake up when the good times come again.

It does sound a good idea on a cold, dark November evening when the clocks have gone back and it is dark by 4.30pm. It will be like this till the end of January at least.

Couldn't we humans hibernate? The problem is if we went to sleep in mid-November and didn't wake up till February, we would wake up dead! We would actually have died of hunger and thirst, and, in truth, wouldn't wake up at all. So how do creatures like the hedgehog manage?

Hedgehogs feed on slugs and snails and dog food if they are lucky enough to be able to raid the dog's dinner in the garden. But in the natural, wild conditions, food for hedge-hogs has vanished by the end of October. The hedgehog seeks out a cosy, safe place deep in piles of leaves, under garden sheds, in old wood piles, wherever appears safe to them, rolls into a ball and goes into a complete sleep until April when rising temperatures wake it up again to face another year. How come the hedgehog doesn't die of hunger and thirst during its long sleep of five months or so?

The trick is that the hedgehog is able to slow its metabolic rate when in hibernation and so its fat reserves last all winter. Normally hedgehogs – which are small animals – have a heartbeat of about 200 beats a minute; they have a body temperature of around 40°C and breathe about 15 times per minute. They need lots of food to maintain this high metabolic rate and so are hungry hunters of snails and slugs, to the delight of the gardeners who are lucky to host them. However, when they are in hibernation their heart-beat reduces to one beat per minute, their temperature drops to 10°C, and they breathe once every three minutes. It takes a lot less fuel to maintain this very low metabolic rate. If they have increased their fat stores so that they weigh at least one kilogram before they go into hibernation and if no nasty event causes them to wake up during winter, they will last until April. Mind you, they will be very hungry when they wake up then and the squashed corpses of hedgehogs that we sometimes see on the roads at the end of April are testimony to desperate forays for food.

We cannot do this. We cannot slow down our metabolic rate so that we barely tick over. We can last a night, but generally most of us are quite happy to have breakfast next day (and two further meals if we can lay our hands on them). A heartbeat of 60 beats a minute and a temperature of 37°C – even when we are asleep – requires a lot of energy, which regular eating provides us with. So, no hibernation for us then.

Bats, which are insect-eating at our latitudes, also hibernate when night-time cold weather reduces the amounts of flying insects upon which they depend. This usually happens around October, and bats then seek out hibernation quarters where the temperature will be steady over the winter. Cellars, crypts and deep caves are perfect. Here they huddle together, hanging upside-down, and hibernate until the warmer weather returns in April. They are all gone by Hallowe'en. So, the very last thing you would encounter when out and about on 31 October are bats – but why should such a scientific fact spoil all the creepy Hallowe'en fun?

Tropical species of bats feed on fruit. They live in tropical forests where there are always trees in fruit or nectar to drink, and so there is no shortage of food and no need to hibernate. Indeed, the fruit doesn't try to escape the way insects do, so they don't need echolocation to hunt it, not like our insect-hunting bats.

And what of the squirrel? Those of us who learn our science from storybooks are convinced that squirrels hibernate. Think of it – is it likely? What do they do in autumn? They gather great supplies of nuts and store them away. The very verb to squirrel something away reflects this. They are good at collecting and storing supplies of food. Therefore, is it likely that, having done all this hard work, they would then go and hibernate for the winter? By the time they would wake up in April, the nuts would have germinated and grown into trees!

Would we go and do huge amounts of food shopping, store it all away in our presses and cupboards and then go off for four months and leave it all behind? Of course not. If we do panic-buy great quantities, it is because we fear food supplies will run out and we will be short. We certainly intend to be around to eat it.

Squirrels are not eejits either. They collect stores of food that will keep because they are going to be awake and eating it over the winter months. They sleep at night in their nests, called dreys, in hollow trees and if it is too cold to go looking for food during the day, they have their supplies to tide them over. They are not obliged to get up and go to work on cold, dreary winter days. Just pop a few nuts in the mouth and turn on the sport on the telly!

We sometimes see squirrels on bright sunny winter days as they are out and about foraging on the ground around their trees. It is easier to see them at this time, when there are no longer leaves on the trees. And they don't forget where they have stored their supply of nuts either, no matter what Enid Blyton says. If they did, survival of the fittest would soon come into play and such dozy squirrels would not survive to pass on such defective memory genes. There are no dig-outs in the world of nature.

In other countries creatures such as dormice, bears and of course groundhogs hibernate for the same reasons – lack of available food.

However, polar bears do not hibernate. They would freeze to death in those icy Arctic wastes if they did. The males go out on the pack ice and keep guard over the breathing holes created by the seals in the ocean below. Seals are mammals and have to come to the surface to breathe. Keeping a breathing hole open during the frozen winters is vital and involves continuous effort on the part of the seal colony. All the bear has to do is sit on the ice beside the hole. Looking up, the seals cannot spot the white mass of the bear against the ice and think that the coast is clear. The first one up for a breath is nabbed, and the rest of the colony can literally breathe easy for another while.

Female polar bears do not hibernate either, although they are not seen out and about. They dig a den in the snow and there they give birth to their young and suckle them for a few months while the weather rages outside. They do not feed at all but depend on their fat reserves to keep going and to provide milk for their cubs. By the time it is warm enough to emerge, they are absolutely starving. Any creature that comes within hunting distance is fair game and that includes humans too.

Mammals are not the only creatures that hibernate to survive the winter. Other groups do it too. In general birds don't, except for a north American species called the Poorwill, which is a member of the nightjar family. As well as being one of the few groups of birds that feed at night – swooping on flying insects from a convenient perch – it also hibernates in winter when its food becomes scarce. It squeezes into holes in trees and goes into a torpor that allows it to survive. Naturally, the Native Americans, on whose lands these birds occurred, knew all about this. But when colonising Europeans, having taken over their lands, turned to studying birds, they were fascinated by the habits of the Poorwill. They carried out great studies and published their discoveries in learned journals. They never consulted their neighbours who always knew about this behaviour, and their native word for the Poorwill was 'the sleeper'. Until important learned white people discover it and publish, it is not 'known'.

Cold-blooded animals don't necessarily have cold blood. They are so called because their body temperatures are regulated by the environment rather than by the energy created by digesting large quantities of food. So cold-blooded creatures in surroundings where it is very cold for part of the year hibernate when it is too cold to function.

Our amphibians – the frogs, newts and Natterjack toads – and our reptiles – the Viviparous lizard and the slow-worm – all hibernate during the winter, when it is too cold for them to move and there is no food available for them as they are all carnivores, depending on insects for food.

Fish don't hibernate – they don't have to, due to the unique physical properties of water. Water is the only liquid known to us that that doesn't continue to get heavier the colder it becomes. When the temperature of water reaches 4°C it is at its heaviest. If the temperature drops further the colder water becomes lighter and rises. So, the coldest water is at the surface when zero degrees is reached and a layer of ice forms.

Your garden pond freezes from the top down rather than from the bottom up, as do lakes and even oceans such as the Arctic. If there is enough space beneath the ice, fish can live perfectly well as the water gets no colder than four degrees. There is more dissolved oxygen in cold water, which they extract through their gills, than there is in warm water, so our fish prefer these conditions to tepid summer water.

Invertebrates – animals without backbones – are all cold-blooded creatures, so they have differing strategies to survive winter. Some do straightforward hibernation.

Snails seal off the undersides of their shells once they have reached the safety of large clumps of overhanging ivy, or some such safe place, and spend the winter there – great groups of them all together if it is a favoured site.

Most of the insects at our latitudes cannot hack it, however, and with the onslaught of winter they all die. Ever wonder where all those wasps that have been plaguing you through September suddenly go in October and November? Even if you find their big papery nest hanging from the roof of your shed or attic it will be completely empty of living wasps.

Bumble bees have all disappeared too, as have bluebottles

and butterflies among others – too cold, too cold! So how do they come back again in spring? They don't migrate, surely?

Wasps, bumblebees and bluebottle flies hibernate only as fertilised queens. They have already safely gone away once mated in August to a secure place to spend the winter fast asleep. Any nice secure place, such as an abandoned mousehole at the bottom of a meadow, or cracks and holes in sheds and buildings, will do fine.

Most butterflies and indeed moths, hibernate in the chrysalis (cocoon) stage of their life cycle and emerge as adults when temperatures rise again. There are a few exceptions – such as the Small tortoiseshell butterfly – that orange one with the blue spots around the bottom of its back wings, which hibernate as adults, sometimes inside our houses, in places like the curtains or in corners of window sills.

Honeybees do not hibernate at all. They were originally a tropical species, coming from a region where hibernation is not necessary. They store great supplies of honey to use as food when flowers full of nectar are in short supply, and they all cluster around their queen to keep her warm. We meanly take some of their honey for ourselves. Their ability to make and store large quantities of honey is the reason why mankind domesticated them in the first place – honey being the earliest source of concentrated sweetness in these latitudes.

So hibernation works as a survival strategy. Is it better than migration? True, there aren't the risks associated with long distance travel, but if you haven't laid down enough stores of fat to tide you through, there is the danger you might never wake up. Or if a very mild spell mid-winter wakes you too soon (climate change and all that), and there is no food available to replenish the energy you used getting your body going again, well, that's not good either.

There must be a better way, one with fewer risks associated with it than either migration or hibernation. And indeed, there is – it is called omnivory.

5

OMNIVORES

The ones that have it all

Migration and hibernation are two strategies that enable many species to survive winter in colder climates, but both come with a great deal of risks, and any particular individual who runs out of luck might not survive until the following spring. And all because the food they survive on is not available in winter. Why put all this effort and bother on themselves? Why not eat something else instead that is available? After all, not every creature hibernates or migrates during winter. How do they manage?

How indeed? What is the story with food and wildlife,

anyway? Well, it is actually quite straightforward. Plants make the food that the animals eat. All the food on Earth is made by the green plants. By the process of photosynthesis in their green leaves, they remove carbon from the air and turn it into sugars in their leaves. Further plant processes make these simple sugars into more complicated molecules, giving us carbohydrates, such as tubers and grains, plant proteins in nuts and seeds, and lipids and fats in vegetable oils. All animal life depends on the food made by the plants – or to give them their full title: the primary producers.

Directly eating plants in order to get enough energy to live seems to be the way to go and many creatures do this. Over the millions of years that life has been on Earth, plants have provided a complete array of food requirements to many groups of animals that have evolved to survive on them. These can range from tiny greenflies, which exist on the sap in leaves, to colossal creatures such as elephants that each need 300kg of vegetable matter per day – a fact that does not endear them to local African farmers if they wander onto their lands. Animals that can live entirely on vegetable material and digest it and get all their energy requirements from it are called herbivores. They range across a wide variety of animal groups.

We have herbivorous birds (such as pigeons, geese and swans), herbivorous mammals (such as deer and wild goats), and herbivorous reptiles (such as tortoises), and the colourful parrotfish of the coral reefs live entirely on the algae they scrape from the coral rocks. Many of our invertebrates are herbivores too, from butterflies and moths to snails and grass-hoppers. Locusts are such voracious herbivores that great swarms of them are dreaded by crop growers in the warm countries where they live.

But each mouthful of plant material has little enough energy, so herbivores spend huge amounts of time eating, digesting and indeed voiding the indigestible remains. Even in dinosaur times, hundreds of millions of years ago, there were many herbivorous species of dinosaurs, whose frequent and abundant droppings have been fossilised, thus conferring them with immortality.

Herbivory is a way of life that works very well for lots of species. Not so well for birds actually, and there are not that many bird species that are herbivores. It makes rearing young more complicated. Birds that are hatched out after fifteen days are blind and lacking in feathers. They must be fed in the nest with material brought there by parents. Vege-table material is low in energy, so great quantities of it would need to be lugged there to provide enough energy for the baby birds, and then there would be the problem of all those droppings.

Pigeons have got round the problem by feeding their

young with milk. Well, it's called milk, although true milk is only produced by mammals in mammary glands. The pigeons produce a milk in their crop – derived from the seeds they eat; it looks like a cheesy mass and is rich in protein. The young in the nest put their beaks into the open mouth of the parent and receive this food directly.

Hummingbirds fly back to their young with their throats full of nectar and squirt it into the mouths of the hungry chicks.

But what if you can't make this milk-like, protein-rich food, or carry mouthfuls of nectar, what then? What do geese and swans do? How can they feed their young? Well they can't – not in the nest as other parents do. What happens here is that it takes up to thirty days for the eggs to hatch out. The young emerge from the egg ready to go. With their eyes open and a protective covering of down, they are fit to walk and leave the nest as soon as their coat has dried. They run about after their parents, following their example and collecting their own food. There is a lot to be said for it.

Herbivory is all very well, but wouldn't it be less hassle if you could get someone else to concentrate the food energy for you, so that you wouldn't have to spend all day eating and excreting? More energy per mouthful – more bang for your buck. That is exactly what carnivores do. Carnivores eat other animals, or at least they dine on the bodies or flesh of other animals that they catch and kill or encounter already dead. There is much more energy in each mouthful of food so carnivores get their energy from smaller quantities of food.

Carnivores can even eat other carnivores, giving rise to a food chain with different levels. Primary carnivores feed on herbivores. These include: lions and cheetahs, who bring down gazelles; birds such as blue tits, who feast on green flies and caterpillars; bats, which dine on night-flying moths; and spiders, who trap small flies in their deadly efficient webs. Some birds, such as sparrows that feed on seeds and other vegetable matter, will resort to catching insects to feed their young as this food is richer in protein.

Other carnivores do not seek a diet sheet from just one species of potential prey but will gobble anything they can catch. Herons, although mainly fish eaters, will devour insect-eating frogs as well as any rat that may be unwary enough to be seized by its rapier-like beak. Buzzards will pounce on and kill herbivorous rabbits, but will also catch unwary fox cubs. Those animals at the top of the food chain are supreme predators. Nothing catches and eats them. Their enemy is the scarcity of prey – especially in the hard times.

No animal can spend more energy finding food than it gets from its meal, so the longer a top predator has to spend catching prey, the more precarious the task becomes. Wolves share the effort by hunting in packs. Other animals let someone else do all the hunting and killing and hang around to grab a free feast. Scavengers such as jackals, vultures and crabs are adept at this, but even our own kites won't turn up their nose at a free meal if they come across one. All these carnivores are only able to get their energy from eating the bodies of other animals. They have all the necessary digestive enzymes to break down and absorb this kind of food. That is why, when perhaps a baby swallow is found fallen from the nest and kind rescuers save it and try to feed it with breadcrumbs, it won't work. The young swallow must have the high-protein, digestible fly food that its parents provide for it. Breadcrumbs won't do, which makes it almost impossible to feed such a rescued young bird, no matter how much the rescuer wants to do so.

Carnivory has its advantages. More energy-rich food means that when times are good the carnivore can have time off to laze about. But when food is not available in winter, hibernation or migration is often the only solution. The best solution of all would be to be able to digest both animal and vegetable material. To have all the right enzymes for both sorts of food. If one type was scarce, well then something else could be enjoyed instead. And, of course, this is what omnivores do. They can eat both animal and vegetable material, and as such are supremely adaptable and abundant.

Robins don't have to migrate to Africa when winter

comes and their supply of insects, spiders and other small invertebrates vanishes. They can feed on berries and seeds. Blackbirds and thrushes change their diets, too, in autumn and avail of all the fruit and berries so abundant then. Many mammals are omnivores. Badgers will feast on blackberries; pine martens are very fond of wild cherries, as proven by great piles of cherrystones in their droppings in early summer. Carp feed on aquatic insects as well as the algal plants around them. Insects such as earwigs and crickets are omnivores, too, which ability, I am sure, must tide them over lean times.

Yes, it is indeed very advantageous to be able to feed on a whole range of different sorts of foods. Being adaptable is the key to survival, as urban foxes or magpies demonstrate. But the most successful omnivore of all − in terms of abundance and distribution anyway − is *Homo sapiens,* ourselves. We have the wherewithal to digest both meat and vegetable matter, and so we have increased and multiplied and spread to the four corners of the Earth. We are the only species that cooks our food, making it much more digestible and this has contributed enormously to our success as a species. We have evolved as omnivores. We have the bacteria required in our guts in order to be able to digest complex starches in our vegetable food. We have flat molar teeth to grind up seeds and grains. We produce hormones, such as insulin, to enable us to get energy from simple sugars such as glucose. But we also have the required apparatus to digest meat. We have pointed canine teeth and flat incisors. We produce enzymes such as pepsin and lipase that break down proteins and fats from animal sources. We cannot make essential vitamins, such as vitamin B12 and vitamin D3, from just a vegetable diet alone. We might not be as good a hunter as top predators, such as tigers, but by cooking our food and farming our food we have been able to increase and multiply.

We do not need the vast quantities of animal products we eat in order to live and thrive. Less than a gram of protein per kilogram of body weight daily would provide us with all our essential nutrients. But we cannot survive as vegans either unless we take supplements to make up for the vitamins we would lack otherwise.

How we satisfy this huge desire for meat is a sorry tale for the rest of the species we share the world with. Killing and eating wild animals and trading them in markets has led to the spread of diseases caused by viruses that have mutated to affect humans in tragic ways, such as Ebola and Covid-19. We use up an unsustainable amount of the planet's resources to produce meat. We have cut down great swathes of rain forest and replaced them with areas of grassland for cattle, thus destroying a huge amount of plant and animal species for which these forests were their habitat. We even break the rules of ecology by forcing herbivores to eat meat so that they will fatten up quicker – at least we did, until mad cow disease put an end to feeding cattle with bone meal and additives derived from ruminant animals, in EU countries anyway.

Herbivory, carnivory and omnivory are ways of eating that have evolved to suit the species concerned and enable them to live in harmony in their habitats. We mess with this at our peril.

6

FARMING

The best idea the human race ever came up with

Humans began farming about ten thousand years ago in the Fertile Crescent around the rivers Euphrates and Tigris, modern-day Iraq, Syria and Lebanon. Here, where there was an abundance of water and fertile soil, people began to grow crops and rear animals. This was a major step, and it eventually led to the end of the hunter-gatherer existence that humans had lived up to then. But there were ecological rules about growing and harvesting crops and animal husbandry, and these soon became apparent to successful farmers.

In a hunter-gatherer existence people gathered what food

they could and moved on. Not every plant was edible, so no clearance of vegetation occurred. Plants died back, decomposed, and renewed the soil so as to allow new plants to grow again – much as we see today in untouched tropical rainforest. A great abundance and variety of plants and their attendant animal species are a continuum of existence. Nutrients are continually recycled between the growing trees and plants and the soil beneath, and the animal droppings and indeed their dead bodies also add to the recycling of nutrients and the fertility of the whole ecosystem.

Farming involves clearing the existing vegetation and growing designated crops. If these crops are continually harvested and removed for food, the soil soon becomes depleted of minerals and nutrients to support future growth. The wise farmer knows that the soil must be replenished with fertilisers to replace that which was depleted and removed with the crop. This can be done with the non-edible parts of the plant harvest – composting in other words – and by letting the droppings of the grazing animals go back into the soil. Even without this, soils can recover if left fallow for a period, and this is what nomadic tribesmen exploited as they moved with their herds seasonally. However, planting unsuitable crops that need more water and nutrients from the soil than are available can be a disaster.

This is what happened in the 1930s in the area that came

to be known as the Dust Bowl in the mid-west of the United States, in states such as Texas, New Mexico, Colorado and Oklahoma. The inexperienced homesteaders moving west to the semi-arid regions of the Great Plains wanted to grow wheat. They ploughed up the established grasslands that were the native vegetation in this dry area and planted thousands of acres of wheat, which did indeed grow for a few years. However, the onset of drought in 1931 caused the wheat crops to fail, exposing the bare, over-ploughed fields. Without the original, deep-rooted prairie grasses to hold the soil in place, it began to blow away. Eroding soil led to massive dust storms and economic devastation. By 1934, thirty-five million acres had been rendered useless for farming and a further 125 million acres was rapidly losing its topsoil. The wrong plants were planted in the wrong place and the natural balanced ecology of the whole area had been destroyed.

The same thing can happen when tropical rain forests are cleared so that grass for cattle can be grown. Surely, one would think, this must be a very fertile area – look at all those huge trees and the luxuriant vegetation. But once the cycle is broken and the trees are cleared, all the nutrients that cycled between the living trees and the soil beneath are gone too. The roots of the forest trees that held the soil in place are also gone and the heavy rainfall causes soil erosion. Even the water cycle, whereby those trees drew in thousands of litres of water through their roots and transpired them as water vapour through the leaves in their canopy has been disrupted, causing adverse changes in rainfall patterns.

You have got to know what you are doing when you are farming. And to be fair, until the Agricultural Revolution of the mid-twentieth century, successful farmers knew not to exceed the carrying capacity of their land. You couldn't take more from the land than it could give. The grass of a goat or a cow were a well-known measure of wealth in earlier times in Ireland and formed part of many a lucky woman's dowry. Keeping up the fertility of the soil was a constant worry to the seaweed gatherers, who collected seaweed for use on their potato patches. A great heap of farmyard manure – a mixture of cow dung and straw – was present in the haggard of every farmer who had tillage and animals.

But then it became possible to fertilise the soil by artificial means. Phosphate, nitrate and potassium could be bought as powders in hundredweight bags, and you no longer needed animal wastes to fertilise the fields. Animals could be fed indoors in slatted sheds over the winter once you could afford to buy in food for them. The link between the carrying capacity of the land and the demands made on it was broken, and the era of intensive farming had begun.

Chicken was a great treat in 1950s Ireland, as housewives were unwilling to slaughter their hens while they were still laying eggs. Nowadays, Irish people are the highest consumers of poultry meat in the EU. We eat so much that our fast-food industry has to be supplied largely with cheap imports from other countries. We eat a lot of Irish-reared chicken too. So, housing chickens indoors and feeding them concentrated diets means that a fast turnaround can be obtained. Broiler, intensively reared, chickens have a lifespan of a mere fifty-six days. A traditional free-range chicken, on the other hand, gets a stay of execution until it is at least eighty-one days old. Chickens running around the farmyard old-style can live up to six years if the fox doesn't get them.

In many countries cattle are intensively reared indoors all the time. Concentrated feeding means that they grow fast and BSE (bovine spongiform encephalitis or mad cow disease) has necessitated that they must now be slaughtered for meat under the age of thirty months so that there is no risk of them catching the disease. In the main our Irish herd feed outdoors on grass during the growing season and are only brought indoors in winter. In fact, 55% of the country is maintained as grassland to support our 6.5 million-strong herd. And I mean grassland – it is managed so that only the best grass species for sustaining cattle grow there. No welcome for wildflowers that are out-competed by the vigorously growing grasses on the nitrogen-enriched soil.

Our 4 million sheep are not intensively reared indoors

but graze all year round outdoors. The numbers are greatly reduced now from the overstocking in the 1980s and early 1990s. Then more than 8 million sheep overgrazed our mountains, causing exposed bare soil, soil erosion and even soil slippages and bog bursts. The carrying capacity of the land was neither here nor there in the economics of maintaining so many sheep, when the EU Common Agricultural Policy (CAP) paid headage on each one.

We are fond of pork and bacon here too. The days when every smallholder had a pig or two are long gone. Most of our pigs are intensively reared indoors, often in small pens, and are slaughtered for the consumer at around six months of age. There are strict EU regulations about the quality of all meat produced and consumed in the EU, and very high standards are set and maintained. Intensive production means that all these meat products can be produced quite cheaply, resulting in edible meat being readily available to everyone.

It is the same with plant food. The more intensively it is grown, the bigger the harvest. Unwanted plants – known as weeds – compete with crops for nutrients and light. Spraying the fields with weed killer, if necessary, before planting, means that the crops get off to a good start. Soil with enough nitrate, phosphate, etc means the crops have no shortage of nutrients. While such an amount of food attracts lots of wildlife to eat it, the shorter life cycles of many insect species means that their numbers can grow quickly and pesticides and insecticides are then deployed to get rid of them.

There are EU regulations controlling the amount of all these chemicals that can be used and specific times laid down to ensure sufficient time between spraying and harvesting. So, in Europe, at least, there haven't been food shortages and famines in quite a long time. The opposite in fact. We ended up with so-called butter mountains, wine lakes, etc when surplus food had to be stored, so much of it was produced – not to mention milk quotas to reduce the amount of milk produced,

All this intensive farming is adversely affecting our natural environment and the creatures in it. Increasing the size of fields, to allow bigger machinery to work there, has meant that many kilometres of hedgerows have been removed and the wildlife for which this was their natural habitat has gone too. The absence of wild flowers in intensively managed grasslands and arable crops has meant that our pollinating insects are starving. Managing fields for silage, rather than leaving the grass to grow to be harvested as hay in late summer, has meant that all the ground-nesting birds that found safety and cover in such places no longer have suitable nesting habitats.

The use of insecticides has greatly reduced the insect population – both the crop-eating ones and more tragically other insects higher up the food chain that would never touch a farmer's crop, such as ladybirds and bumblebees. Bats and aerial bird feeders, such as swallows, swifts and house martins, not to mention cuckoos, who just love hairy caterpillars, are also facing famine. Cattle dosed with medicines for intestinal parasites excrete some of it in their droppings. This causes the dung beetles that used to break down this vital natural source of soil fertility to be greatly reduced in numbers as well, and black, undecomposed cowpats litter the fields the grazing animals use in summer. If a field is given more fertiliser than can be absorbed by the growing plants, whether that is artificial fertiliser or even a layer of slurry spread with a dung spreader, then our abundant rainfall will wash it off into our waterbodies, our rivers and lakes. This causes overgrowths of the algae that are naturally in all our fresh waterbodies, thus depleting oxygen levels at night when the green algae do not photosynthesise. It will also increase the bacterial content of the water if enough organic waste such as slurry gets washed in. It is a real problem for farmers if they have more animal wastes because of intensive rearing than can be absorbed safely on their own land – in other words if the carrying capacity is exceeded and it can affect their animal stocking rates.

Again, there are standards and regulation set by the EU for all this. Much of Ireland's soil has been designated as having high-nitrate or high-phosphate levels and so there are limits to the amount of fertiliser or slurries that can be spread on them. Our lakes and rivers are monitored for water quality and while we have very little truly bad water quality in our rivers and lakes, the amount of pristine water at the very highest standards of excellence is diminishing year on year.

But is our food safe to eat? Of course it is – that is why all these standards have been drawn up and enforced. What the farmer produces has to reach acceptable standards before it is offered to the consumer and it does. Intensification may have changed the taste of much of our food – or at least high-yielding, fast-growing varieties are grown which do not have the taste that slow-growing traditional varieties have. Muscles that are used by animals that can move about result in meat with much more taste than that from animals intensively reared in small spaces.

Organic farming is farming that uses no artificially pro-duced chemicals of any sort and allows animals free move-ment. Soil is managed in such a way that the natural fauna such as earthworms manage its fertility. No pesticides or weedkillers are used on organic farms and a greater vari-ety of crop species are grown. But it is not easy. There is much more labour involved in manually removing weeds from around your crop plants. No insecticide use means that creatures that want to eat your plants must be deterred by physical means such as nets or scarecrows or eaten as part of the food chain by all the natural wildlife in the hedges and trees on your farm. Barn owls and buzzards take care of any rats, which, not having consumed rodenticides, do not harm the birds of prey. Chickens have freedom of movement, and as well as being fed food with no chemicals are also able to scratch around in the soil for extra titbits. Their eggs have more flavour, and their muscles are well developed. But there is always an ongoing battle with the fox who likes free range birds too.

So, an organic farmer's crops are hard-earned and the yield is lower when all this is taken into consideration. No wonder it costs more when it is offered for sale.

We eat much more processed food nowadays than ever before. In the days when one parent – usually the mother – stayed at home, meals were home cooked from raw materials bought in shops and markets or home grown. Soup making, bread baking and making cakes were all everyday chores and the cook knew exactly what went into the meals that were prepared. It all took time, which busier generations don't feel that they have nowadays, to shop and cook meals from scratch. Processed jars and packets make preparing meals at home faster and take-away food means that it doesn't have to be prepared at home at all.

As a consequence the diner can be at the mercy of what the manufacturer has put into the products. The ingredients must be listed on the packaging and indeed they are. Higher levels of salt and sugar enhance the flavour of the food. Other ingredients are there to prolong the shelf life – because there must be dates on the packets to say when they should be eaten by. Again, all these ingredients must be approved, and the product must be safe to eat, but there seem to be far more ingredients in a shop-bought cake than ever went in to the home-made sponge. And the food tastes different too. But no matter, if it is quick and handy you can be tucking in to your dinner in time to watch the latest cookery programme on the television.

We worry if too much sugar is bad for us, or too much salt? Should we change to low-fat food? Should we use honey instead of sugar, would that be better? On that score, the grains of sugar we use are actually sucrose. The sugars in sweet fruit are fructose and glucose while that in honey is mainly fructose. Lactose is the sugar in milk. They are not good or bad unless you are diabetic and lack insulin to break them down. It is the amount of them you eat in your food, rather than the type, that can affect your health.

We worry about allergies to and intolerance of certain foodstuffs. We speak of people being lactose intolerant as if this was somehow an aberration. What actually is an aberration is being able to digest the milk of another mammal species such as cows, after the age of five or six. Only white

Caucasian people can do this and indeed only those that have the mutation that allows them to do it – which in fairness is most of them. The rest of the world's peoples in the main do not have this mutation and so cannot digest milk as adults.

Having an intolerance of gluten – being coeliac – is another mutation. It is common enough among Irish people. If a mutation in a race of people is a disadvantage, then it will not be prevalent. But in a country where a huge population once depended on potatoes – which have no gluten – as their staple food, being coeliac didn't matter in the same way as it would if your basic food was made from wheat or rye. Mind you, the changes in the variety of wheat being grown can affect how people digest it. Plant breeding has developed strains of wheat that have shorter stems to avoid lodging during wet weather, or ones with higher-yielding ears compared to old-fashioned strains such as spelt. But being coeliac means that the body cannot process the protein gluten and it is an auto-immune disease. I suppose, really, if something doesn't agree with you it makes sense not to eat it. In our first-world economy there is always something else to eat.

We have more control over what we eat if we cook our meals ourselves. When did cooking become such a chore? Why do so many people proclaim that they cannot cook? Surely if you can read, you can cook. Processed food is designed so that it will have a good taste – the manufacturers put a lot of research into it. They can't actually poison us you know. All food must reach acceptable standards and constant inspection ensures that this is so. It is up to us to decide what we want to eat and how much time we want to devote to preparing our food.

Farming has ensured that, in the main, there is stability in food production. Such food security has been one of the main reasons why, after growing exceedingly slowly for more than half a million years, the world's population has greatly increased over the past 10,000 years. It is estimated that 10,000 years ago the world's population was less than 2.5 million. It is now almost eight billion. That is a lot of mouths to feed.

7

WHY DO BIRDS FLY?

Because they can or because they have to?

Since the time of Daedalus, way back in the mists of mythology, people have been fascinated with birds and how it is that they are able to fly. Daedalus, the master craftsman, made wings for himself and his son Icarus from feathers and wax. They used them to fly away from captivity in Crete as they could leave by no other route. It worked well and they succeeded in escaping. Daedalus had warned his son not to fly too high because the fierce heat of the sun would melt the wax that held the feathers on, but of course the young fella got carried away and flew higher and higher. And of course the wax melted, the wings disintegrated, and Icarus fell into the sea and was drowned. That taught him to do what his daddy had expressly told him not to.

In fact, the story bears no scientific scrutiny whatsoever. The higher you go, the colder it gets – otherwise why is there permanent snow on the top of Kilimanjaro which is practically on the equator in Tanzania? And wings made from feathers do not guarantee that the wearer will be able to fly just by having them. Bats can fly and they have no feathers at all on their wings.

Birds are heavier than air. Why don't they fall down? The question greatly intrigued Leonardo da Vinci, and he published a whole codex on the subject in 1505. But, alas, none of the flight machines he invented actually flew. Nowadays, we board huge enormous airplanes (or at least we used to until recently) that are loaded with fuel and passengers and luggage and tear off down the runway and up into the air. Does it ever cross our minds to wonder how this great thing gets up into the air and stays there? Or do we feel in the moment of take-off, as noted, albeit not very scientifically, in the James Bond film, *Diamonds Are Forever,* 'If God had intended man to fly, he would have given him wings.'

So how did it all happen? The first bird, as we know from fossil history, was called Archaeopteryx and lived 150 million years ago. What enabled scientists to declare it to be a bird was the fact that it had feathers. It had reptilian features too and was described as the missing link that proved that birds evolved from reptiles. Being able to fly was not considered to be an essential prerequisite to be a bird. Archaeopteryx could at the most glide for a certain distance but not fly, the same as several other groups of reptiles such as the Pterodactyls, which were also around at the same time.

True flight meant that the bird was able to take off – like our modern-day airplane – and then propel itself through the air, its flight muscles providing a continuous source of energy to keep it going. This is much like the jet engine or the propellers on a plane that are a continuous source of energy, propelling forward motion. Flying backwards is not an option. By the Eocene epoch, 54 million years ago, the birds had cracked it, and the fossils from this period show evidence of a well-developed keel on the breastbone where very strong muscles were attached that could sustain flight.

Flying requires huge amounts of energy on the part of the bird. The whole body has evolved to suit this. To make them as light as possible, the bones are hollow but are strengthened by a honeycomb of struts. Other bones are fused together forming rigid girders in the skull, pelvis, wings and back. They have a supremely flexible neck with as many as twenty-five vertebrae in all, allowing it to bend and twist freely. Mammals – even giraffes – have only seven vertebrate, not much head-twisting there.

But it is the breastbone that gives the game away. This is a plate of bone with a keel jutting out, to which the massive flight muscles are anchored on both sides, as any cursory examination of the roast chicken lying on its back on the carving board will illustrate. Were you to have an albatross lying there instead of your chicken, the keel would certainly impress you, with its size and rigidity. Wing feathers are essential for flight, there being five different groups all working together as the wing propels the bird through the air. Trimming the primary wing feathers is enough to prevent captive birds from flying away, although their keepers have to be vigilant as they will grow back whenever the bird moults and grows new feathers.

A bird must be streamlined all over, so its reproductive equipment is internal in both males and females. Eggs are formed and laid one at a time, and the offspring develop externally in the laid egg. No female bird can afford to look pregnant.

And why do birds fly? Because they can. The air is a whole wonderful habitat, full of food, free from mammal predators, where they can make a living. Once they conquered flight 54 million years ago, there was a great species explosion, as all sorts of ways of adapting to and exploiting this uninhabited world evolved.

But flight takes a huge amount of energy. There can be good food on the ground – aerial insects can be overrated! It's the predators down there that steal your young and eat the eggs that are the problem – those pesky mammals, snakes, etc. If you could find a place where they weren't around, then you could ease off on the effort required to fly. And that is what many bird species did. They abandoned flying in safe places and soon lost the ability to do it. Use it or lose it, and they did lose it.

Large islands that those Johnny-come-latelies, the mammals, never reached were good places to live a less energetic life. Australia and New Zealand had broken away from the rest of the world before placental mammals, such as rats, evolved and long before humans evolved. Large islands such as Mauritius were too far from continents that had placental mammals to be colonised. Places like the Antarctic and the Arctic were too cold for predatory land mammals interested in eating birds' eggs. In these places it was safe to abandon the terribly energetic activity that flying was and have an easier life style.

And you could grow bigger and heavier, too, if you didn't have to launch yourself up in the air. The heaviest bird ever was the Elephant bird, which was once a common sight on Madagascar. These birds weighed at least 500kg (half a tonne), stood at around 3m tall and laid giant eggs. The birds on New

Zealand were no titches either. There were nine species of moa there, the two largest of which reached about 3.6m in height with neck outstretched and weighed about 230kg (a quarter of a tonne). The ostrich – our biggest extant bird today – is but a mere 2.7m in height and weighs in at 145kg.

If your preferred food was fish, why spend all that energy flying in the air when your food was swimming underwater? The Great auk was a large flightless bird that lived in the polar regions of the northern hemisphere. Penguins – all of whom are native to the southern hemisphere – manage very well to this day without having to launch themselves into the air.

It is a good way of living, or at least it was, until the deadliest mammal of them all took an interest in flightless birds – us humans. Many of the flightless bird species lived in parts of the world uninhabited by humans. When humans suddenly appeared in their territory, they had no reason to feel threatened. But man – the supreme pack hunter – saw flightless birds as an easy source of food.

Hunting pressure from humans and habitat destruction of the forests where these birds lived meant that the elephant birds were all gone by 1500 years ago. Remains of bones found show marks of butchery, proving that they were killed and chopped up by creatures with sharp cutting tools – humans, in other words.

The same fate befell the moa birds in New Zealand. The first human arrivals there from further northwest in the Pacific found a lush set of islands, on which no mammals existed. They settled there at the end of the thirteenth century, and just a mere two hundred years later had caused the extinction of all nine species of moa. The poor devils hadn't a chance, as eggs, young and fully grown adults were all targeted as food.

There are still some flightless species of bird left in New Zealand. There are several species of kiwi, which survived by hunting at night for food under the leaf litter on the forest floor. Mice and rats brought, probably inadvertently, by the humans are giving them a run for their money now as they outcompete them and indeed the rats raid their nests and eat their eggs. There are five species of kiwi and other flightless birds, such as the kakapo parrot and the henlike takahe, and they are being protected – not always successfully – by massive conservation efforts.

Being able to run faster than the enemy is a good way to protect yourself and this is the reason why our existing large flightless birds have survived. Living in open grassland and having great eyesight also helps.

In South America it's the rhea, in Africa it's the ostrich and in Australia it's the emu who all have made the grassland areas their territory. Having strong leg muscles helps too, and a kicking ostrich can see off any unarmed predator.

Penguins have survived in the southern hemisphere, where the many different species can breed on the frozen vastness of Antarctica far away from human predators. The northern flightless species, the Great auk, wasn't so lucky. They were able to live in such cold places because they were well insulated with down. Nesting colonies on the European side of the Atlantic were nearly all eliminated by the mid-sixteenth century by collectors of down for use in pillows. On the American side, hunters of the Eider-duck for down switched to Great auk in the late eighteenth century, when the eider became increasingly scarce. Specimen collectors then realised how rare they were becoming and sought to obtain one or two, including their eggs, for their specimen collections. And so it was that in Iceland in 1844 the last known pair were killed by collectors, who crushed the birds' egg underfoot for good measure as they strangled them.

And yet we persist in calling birds who became extinct stupid. The dodo is a byword for someone who is an old fuddy-duddy stuck in their old-fashioned ways. Let's hear it for the dodo! It lived happily on the island of Mauritius in dense woodland, minding its own business, until it was first spotted by Portuguese sailors in 1507. By 1681 the dodo were extinct, their forest habitat destroyed and their eggs preyed on by mammals introduced by the humans. Why are dodos considered to be the stupid ones in this scenario? I suppose because history is written by the victors.

So why do birds fly? They fly because it has proved a very successful technique for survival for the past 54 million years at least.

8

THINGS THAT GO BUMP
IN THE NIGHT

Wildlife in the dark

It is getting dark. The dusk chorus is finally winding up and even the blackbird is going to bed. You have finished the last of your sundowner and are heading indoors. The garden is closed for the night. Except it isn't. Gardens are hotbeds of activity on warm summer nights, and indeed on wet autumn nights too. Completely different creatures appear and treat it as their own. These creatures prefer the night; it's cooler, darker, damper, and there's no drying fears from a hot sun. Fewer enemies too.

If, like humans, sight is your most dominant sense, then you can't operate too well in the absence of light. How can you see to find your food? Or a mate? Most birds operate by daylight and are most appreciative of it. The dawn chorus just before sunrise on spring and summer mornings is territory-owning assertiveness, carried out by testosterone-filled males, until it is bright enough to go looking for food. The chorus fades then, as singing on an empty stomach can only go on for so long. In their forest habitats in Africa, chimpanzees and gorillas build nests in trees once darkness is about to descend and stay there until it is bright enough to start looking for food again. Butterflies and bees fly around during the day, seeking nice open flowers with supplies of nectar, to keep them going.

Once darkness falls, however, creatures that do not depend on sight come in to their own. Well-developed senses of hearing, smell or touch mean that life can exist perfectly well in the absence of sunlight. Chief among the animals that are fully adapted to existing in the wild are bats and birds such as owls. But just because one sense is highly developed, that doesn't mean that other senses are necessarily diminished. Bats are not blind, no matter how often humans unscientifically accuse each other of being as blind as a bat. How this ever became an expression is a mystery. Any simple experiment would demonstrate the falseness of it. Suppose you took two people to a dark wood in the middle of the night and asked them to run as fast as they could through the woods. They would crash into the trees and fall over dead branches and generally have a terrible time. Substitute two bats for the humans and let them off and they would wheel between the trees and branches, not only not crashing into anything, but also catching moths, midges or mosquitoes for their supper as they went. Supposing an observer from Mars, or somewhere, was looking down observing all this carry-on. Who would they say were blind? The bats or the humans?

Bats are highly evolved to operate in a world completely devoid of light. The insect-eating bats that live at our latitudes navigate using echolocation. Each bat species emits hunting calls at a particular frequency for that species. Soprano pipistrelles emit sounds at 55kHz, while the larger Leisler's bats emit calls at around a much lower frequency of 25kHz. When these sound waves hit an object nearby, they are reflected back at a slower speed. As sound waves slow, the pitch drops. The bat, hearing the reflected sounds, can determine the drop in pitch and ascertain where the object that deflected the sound is. In this way, it builds up an acoustic image of its surroundings. So, it is highly unlikely to fly into your hair – the sound waves it emits bounce off your skull and back to the bat. Unless your hair is full of moths, midges or mosquitoes, why would it bother?

Bats can fly in complete darkness at great speeds using this technique. They are well suited to come out and forage on the wealth of insects that fly at night – safe from competition from insect-eating birds such as swallows and swifts that fly by day. Bats such as Daubenton's bats fly over water bodies, where many insects congregate on warm summer dusks. Lesser horseshoe bats, being able to hang perpendicularly from surfaces without needing any support from walls, can roost in caves and live in the limestone regions of the west of Ireland, where they hunt from dusk to dawn in summer.

Owls detect their prey by sound too, but they operate in a completely different way. We have two native species of owl in Ireland – the Barn owl and the Long-eared owl. The latter got its name because it has what seemed to observers to be two large ears sticking up straight from the top of its head, but these are not ears at all; they are merely elongated head feathers. No, the hearing apparatus of these owls is much more sophisticated than this. Like all birds, they detect where a sound is coming from by assessing the time lag in the arrival of the sound in one ear and then the other. Owls have their ears placed asymmetrically on their heads, which increases the time lag, and the broad round discs of their faces act like satellite dishes to receive the faintest sounds. Unsuspecting mice and rats never know what hit them as the birds fly in complete silence, having feathers right down to their toes.

Many insect species fly at night. Moths, for example, are synonymous with night flying, although there are a few day-flying ones too. Moths depend on their sense of smell to enhance their nightlife. The reason they are out at night is to visit singles' bars, looking for mates. Like butterflies, moths do not eat as adults. All the eating is done as a caterpillar, which is the eating part of the lifecycle. When they have eaten and burst four times, getting a hairier skin at each stage, they go away from the food plant, become a chrysalis and emerge as a fully formed adult with wings. They have no stomachs and no guts; they no longer eat – all their energy comes from the food they ate as caterpillars. But they can drink, and what they seek is sweet nectar from flowers. How do they find flowers in the dark of night? They smell them, that's how. Certain flowers, like honeysuckle, have long narrow entrances into their blooms. They need to attract night-flying moths for pollination. In order to do this, the flowers only produce perfume when darkness falls.

Having duly tanked up at these conveniently advertised drinking establishments, they then seek out mates. Some male moths can detect the pheromones of a female moth up to two kilometres away, so sensitive are the detection cells in their feathery antennae. I presume he sees her when he gets there.

If you haven't emptied the saucers of water under your flower pots for some time, or you have a pond, you may indeed have habitat for mosquitoes. We have twenty different species in Ireland, although none of them carry malaria here … at the moment. They did carry malaria here long ago when the weather was warmer. Malaria was referred to as the ague, and the last person of note to die from it was Oliver Cromwell. He died in 1658 of complications from malaria that he picked up on his Irish campaigns. The weather got colder after that for several hundred years and while the mosquitoes could still survive here, they could no longer carry the malaria-causing organism.

Be that as it may, female mosquitoes are the ones that lay eggs and they look for a nice warm mammal to bite, so that the feed of blood sustains them while they do this. How do they detect such sources of nourishment in complete darkness? When mammals – including ourselves – exhale, there is more carbon dioxide in our exhaled breath than in the general air around us. Mosquitoes are able to detect this enhanced level of CO_2 from a distance of up to fifty metres away. If this is us asleep in our beds with the window open, then in they come and now our skin odour attracts them. If they prefer yours to that of your sleeping companion, then it's you that they hone in on for the feast. Then, off they go to lay their eggs in the nearest still water, and the cycle continues.

Feeling your way around in the dark is another way of getting out and about. The whiskers on rats and mice aren't just for show. They are equipped with sensory cells at the tips to enable their owners find their way without light to guide them. Not much street lighting in city sewers, as you might imagine. Whiskers can detect changes in air currents as well, so that their owners don't run bang smack into a wall while running in the dark.

Of course if you can make your own light you can always flash a Morse code, as it were, when you wish to attract attention to yourself. Although we don't have these in Ireland, these notice-boxes of the insect world include glow worms and fireflies. These are both species of beetles, neither of which have definitely been recorded in Ireland. It is the flightless female glow worm that can produce a light on the underside of her last three abdominal segments. These segments contain the wonderfully named luciferin which can be catalysed by an enzyme to produce this light. The female makes the light to attract the flying lightless male when she feels in the humour for mating. She can switch the light off too if she is disturbed or I suppose if she changes her mind. Fireflies, which live in warmer countries than either Britain or Ireland, make a cold green light in the same way. Again the main purpose is the same as that of the red lights in certain squalid areas of old port cities, but there are some crafty females that imitate the flashing light sequence of species other than her own and in this way lure in a nice tasty meal instead. It's all go out there in the world of darkness.

It is generally much quieter at night than during the bustling daylight hours – not least because noisy humans are not so abundant then, so some species take advantage of this to call out for breeding partners or indeed to warn other males off their territory. The corncrake call has long been a familiar summer night-time call in rural Ireland. It has dwindled practically to nothing as the corncrake population has declined to just a few hundred calling pairs. The corncrake is found more abundantly in Eastern Europe, in places where traditional agricultural methods are still carried out. In Ireland, our few remaining corncrakes tend to turn up faithfully every summer on Tory Island off Donegal or the Mullet Peninsula in County Mayo, places where a more traditional country life is still to be found.

The nightingale is another migrant species that transforms the night with its melodious song, but not alas in Ireland. It inspired poets such as John Keats and William Wordsworth who were familiar with it in the English countryside, though it is most unlikely ever to have sung in central London in Berkeley Square, no matter how much Vera Lynn or Bing Crosby tried to convince us. But we could occasionally, if we were lucky, hear the song of the wonderfully named nightjar, or even more evocatively titled *tuirne lín* in Irish. This name means spinning wheel, and the churring song, rising and falling in pitch and continuing for a long time, reminded those who heard it of the sounds of the spinning wheel. The spinning wheel, once part of the furniture in prosperous cottages up and down the country, is no longer with us, nor indeed are many calling nightjars. They seldom breed here now, although they were once a widespread summer visitor.

Not so melodious is the call of the barn owl, which many a scared listener has confused with that of the banshee – a much-feared fairy woman who would appear when someone was going to die. But, again, it is a communication call from a species that sleeps by day and so needs to do all its business during the hours of darkness. It's a busy world out there whether there is light or not.

9

IT'S A WONDERFUL WORLD

You just need to know where to look

If you are not familiar with common and widespread plant and animal species, then it is very difficult to appreciate those that are rare and perhaps endangered and threatened as well. And it can be hard to understand why great efforts must be made to conserve them. This first became apparent to me in the 1980s while taking trainee primary school teachers on nature walks on the Aran Islands. This was a decade before Environmental and Scientific Studies became part of the primary school curriculum, so learning about our wonderful wild world had not been part of their education. How could they be expected to go into raptures of ecstasy upon seeing choughs, those members of the crow family with brilliant red bills and scarlet legs, when they were not familiar with the differences between common crows such as rooks and jackdaws? Or how could they have known that habitat for these choughs was quite rare in Ireland and was coming under more pressure from changing land use?

So, when I got the opportunity to present a whole series of wildlife episodes for the then enormously popular children's programme *The Den*, at the end of the 1990s, I made sure that the common residents of my back garden were the stars. These were five-minute weekly slots on such interesting creatures as bumblebees, bluebottles, snails, centipedes, woodlice, etc, all of whom performed splendidly once the cameras started rolling. Ladybirds obligingly fell into my upturned umbrella when I shook the leaves on the tree. Spiders – great big hunting ones – were always in my pitfall traps on inspection in the morning – or so it appeared to the viewer. And as the episodes of *Creature Feature* were shown on repeat for the following five years, a whole generation of younger children – not to mention university students lounging at home looking at afternoon telly in those pre-internet and laptop days – became aware of and interested in what could be found just outdoors wherever they were.

The ability to notice things and the curiosity to ask

questions are the marks of a scientist, no matter what age they are or live in. And this was what I was hoping to inculcate in the viewers. I still meet people who remember those episodes!

I got a second chance to spread the word about wildlife ten years later. This time I was doing my usual introduction to an annual summer course on Environmental Studies. These summer courses generally take place every July as part of continuous professional development for teachers. I was bemoaning what seemed to me to be the lack of knowledge in primary school pupils about the wildlife that surrounds them in their everyday life at home and at school. Baby infants, I declared, come to school knowing three flowers. They know about daisies, dandelions and buttercups. The more advanced ones know that if a buttercup shines under your chin you love butter and that touching dandelions causes you to wet the bed. But when sixth-class pupils – who have been in school for eight years – are asked what wildlife they know, they trot out the same three plants and the same stories. I confidently told the teachers that if the pupils had learned about two plants, one tree, one bird, one mammal and one creepy crawly each year – just ordinary ones all around them – they would know all about forty-eight common local wildlife species by the time they left primary school.

The Laois Heritage Officer who was hosting the course challenged me. Why didn't I do something about it? She would get me the funding and I could write the book of the forty-eight common species that everyone should know. And so I did – putting it on immortal paper this time, not on any technological products that could be overtaken by time. The book, *Wild Things at School*, was for the teachers – still is and is now available online for free – and it enabled them to use their ordinary school surroundings as a wildlife habitat, easily accessible and endlessly interesting throughout the year.

Because that's the thing about education. You lead from the known to the unknown. Knowing what is generally around us, gives us a baseline against which to compare extremely rich and undisturbed habitats and to value them. If you were to believe many of the wildlife documentaries on television, you would think that these have to be the tropical rainforests or the underwater slopes of the warm coral reefs. But not so. On a trip to the wonderful Białowieza Forest in eastern Poland, our very experienced and knowledgeable guide confessed that he would give anything to visit our Irish breeding seabird colonies in summer as there were no such things in Poland. And when my friend on her honeymoon in southern Argentina was taken with great reverence to visit a bog – the only bog they said in the southern hemisphere no less

— she was underwhelmed. In fact, she felt she wouldn't cross the road at home to look at it as it was a very poor example of a bog compared to what she was used to in Ireland. Yet, do we really appreciate the wonderful habitats our bogs are? Do we actually know that there are two sorts — raised and blanket, and could we even tell the difference between them?

Faraway hills are not greener. When I was young, I knew for a fact that faraway hills were not green at all — they were blue. We lived on the top of the watershed between the rivers Dee and Glyde in County Louth and had an unimpeded view across the flat plain of north County Louth to Slieve Gullion, the Cooley Mountains and the Mournes. And they were always blue — never green. It was only when around the age of eight or nine I was brought with my father on his annual summer holiday back to southwest Donegal and was actually among the hills and mountains that I saw that they were not blue at all but green, and brown, and purple and gold, depending on the vegetation on these blanket bog-covered slopes. (Neither indeed were the mountains of north County Louth actually blue when I went to climb them years later.)

Those Donegal hills were completely different habitats to the rich farmland of mid-Louth, and even at that early age I was fascinated. The crows in the sky were ravens that croaked rather than cawed like the County Louth rooks; there were wonderful plants that actually ate insects rather than the other way round; and you could bounce up and down on the ground, which was quaking water-filled peat. The sky was filled with skylarks – and midges, when the wind dropped. A totally different world.

The bogs covering many of our mountains are blanket bogs. These began to form about five thousand years ago when our climate got wetter and colder than it had been formerly. These cold, wet conditions facilitated the growth of Sphagnum moss, which thrives under wet conditions and holds lots of water in its leaves. Gradually the mountainsides that, up to this, had been covered in pine forests became waterlogged. The trees could no longer grow and died where they stood. But lack of oxygen in the waterlogged sphagnum-covered soils meant that no decomposing organisms such as bacteria or fungi could survive. Nothing broke down or rotted away. Layers and layers of moss grew and died, with the dead tree roots trapped in them. Over the following five millennia these became compressed by their own weight, giving us peat or turf with what became known as bog deal (the old pine roots) interspersed throughout.

There are still wonderful examples of blanket bog around today. Go to west Galway and walk on the great expanse of Roundstone Bog, considered to be without parallel anywhere nearer than the Outer Hebrides in Scotland. Take the

R341 off the main road to Clifden (N59), west of Recess, going towards Roundstone, and it brings you on to this wonderful wild region, fortunately saved from having an airport sited there – a plan first mooted in the heady days of 1989 and doggedly resisted over the following ten years. Even from the public roads you get a sense of the uniqueness of the place.

But maybe it's a raised bog you are interested in seeing. Then you must make your way to the midlands of Ireland where there is a different landscape altogether. Raised bogs are older than blanket bogs. They began to form after the end of the last Ice Age, 10,000 years ago. Great lakes of water were left behind in low-lying districts when all the ice covering Ireland had melted. These lakes gradually filled in with vegetation over the millennia. The vegetation did not break down or rot away because of the lack of oxygen and the increasing acidity, and so it all turned to peat. These bogs were often in areas of good land and people always lived nearby. They realised that things would not rot away in the bog, so they often used them as storage places for food or for hiding treasures or, indeed, for drowning enemies in. Turf cutting over the years uncovered many of these, as a visit to the National Museum of Ireland, where many of these treasures are exhibited, will explain.

These raised bogs could contain up to thirty metres of

turf, depending on the depth of the original lake. They were a source of indigenous fuel in a country that had no coal deposits and depleted woodlands. Bord na Móna was set up in 1946 to develop Ireland's peat resources for the economic benefit of Ireland. As well as providing fuel, the turf was also used to generate electricity and to supply gardeners with peat moss. As a result, the raised bogs are mostly cut away now. The best of what is left are conserved as Special Areas of Conservation, but we really have no untouched raised bogs left. Clara Bog in Offaly is worth a visit. There is a visitors' centre in Clara, and then you can go walking on the bog itself and feast your senses on creatures such as frogs and lizards, snipe and maybe merlin, if you are lucky, not to speak of the wealth of butterflies and dragonflies that make the raised bog their home.

But what of the wonderful seabird breeding sites that our guide in Poland was longing to see? He was not wrong. Ireland has seabird breeding sites of international importance. If you ever manage to book a slot to visit Skellig Michael off the County Kerry coast, your boat will take you past the Little Skellig, breeding grounds for the second largest colony of gannets in the world, with over 30,000 breeding pairs. You don't land, but passing it on an open half-decker is the most unforgettable experience. The air is white with thousands of gannets coming and going. The noise is at crescendo level and the smell is absolutely appalling. Gannets build nests from seaweed that rots. They feed their noisy young regurgitated fish, all of which doesn't go down the hungry throat, and the guano deposits from so many birds have built up over the years. Suddenly you very much regret the full Irish breakfast you ate so happily an hour earlier, but the gannets are pleased with the unexpected bonus projected over the side of the boat.

There are only six breeding colonies of gannets in Ireland – all on offshore islands. Great Saltee off Wexford and Lambay off Dublin are privately owned. The sea stack colony off Clare Island in County Mayo and Bull Rock off County Cork are inaccessible too, but you can land on Ireland's Eye, the island just north of Howth and walk to where the nests are. Keeping your distance from the gannets is recommended.

The Cliffs of Moher off the County Clare coast attract a fair share of visitors to see the spectacular scenery. This is also one of Ireland's most important seabird colonies. Here can be spotted fulmars, shags, kittiwakes, guillemots and razorbills – all colonial breeders, but each species with its own niche on the cliff face. June is the time to visit, if it's seabirds you are interested in.

The seabird colonies around the lighthouse on Rathlin Island, off the coast of County Antrim, provide very easy viewing as the thronged breeding grounds are overlooked by a viewing platform above, manned by very helpful RSPB staff.

Some seabirds are so adapted to life at sea that they do very poorly on land. But they must come ashore to breed, so this is a conundrum. They have solved it by nesting in burrows underground and coming ashore in darkness to evade the marauding gulls that would prey on them. The most commonly seen of this group is the puffin, a bird everyone recognises. It nests in abandoned rabbit holes on offshore islands, preferably ones free from American mink and rats. There are plenty of puffins to be found on the Great Skellig, if you ever manage to get out there. The Manx shearwater also nests in burrows on islands off the Kerry coast, such as those of the Blaskets. The Storm petrel comes ashore too in the hours of darkness to feed its young or to relieve its mate of incubation duties. They much prefer old stone walls of long-abandoned dwellings. Being in the presence of these seabirds is not for the faint-hearted. Usually camping overnight on an uninhabited island is required. Once darkness has descended, around midnight or so, the air is filled with the unearthly cries of petrels or shearwaters, calling to locate their mate or chick in absolute darkness. It is hard, if not impossible, to see them. With all the eerie, unending screeches and calls, the master storyteller of mystery and the macabre Edgar Allen

Poe would be in his element.

Rockabill Island, off the north Dublin coast, holds one of the most important tern seabird colonies in Europe. Terns are elegant white summer visitors with forked tails, well deserving of their nickname 'sea swallows'. Five different species breed in Ireland in summer. They are all ground-nesting birds and prefer deserted stony beaches or machair-type sandy areas. Three of the five species of terns – Arctic, Common and Roseate – nest on Rockabill. More than 80% of the total European population of Roseate terns breed here, so the island is carefully wardened to make sure they are free from interruption. The adults feed on small fish and can be seen diving for food all along the coast.

If you know where to look, Ireland is spectacular for birds in winter too. Relative to countries further north, our mild temperate winters mean that we get great flocks of winter visitors who come here for the food.

At peak times in winter, the Wexford Slobs, just north of Wexford town, host up to 10,000 Greenland white-fronted geese – more than a third of the world population. The area is managed for these visitors from Greenland, so that these vegetarian birds – herbivores – have a range of food, ranging from cereal stubble fields on arrival in October, through short grass fields to special planted fodder beet plots in mid-winter. Then, back to the newly emerging spring grass in March to fatten up for the long haul back to Arctic breeding grounds. A visit there anytime during winter is well worthwhile. Not only will you see these geese, you will also see other winter-visiting geese too. Most of the pale-bellied Brent geese in the world spend the winter in Ireland, feeding on grass and other vegetation, and they are easy to identify, being our smallest goose.

There are great numbers of them on Bull Island – that wonderful place in Dublin Bay. It was only created two hundred years ago when Captain Bligh – he of Mutiny on the *Bounty* fame – had a wall built along the northern part of the exit of the River Liffey into Dublin Bay. This together with the South Bull Wall, which had been built earlier, increased the flow of the Liffey into the Bay and stopped the dangerous silting that was such a hazard to shipping. The silt had to go somewhere and where it went was northwards behind the North Bull Wall, forming an island – Bull Island. To the east of this long north–south strip were sand dunes and on the western inner side, where the currents carrying the silt were slower, there developed mud flats and saltmarshes. It was, and still is, a magnificent wildlife habitat – in both winter and summer, for different reasons.

As well as hosting the Brent geese, the mud flats are a paradise for those other winter visitors – the thousands of waders that fly here from Russia and Scandinavia to probe the soft, unfrozen muds for a whole range of invertebrates that live there. Lots of species can exist together, without competition, as depending on how long your legs and bill are, you can source food at different levels and in different depths of water.

Waders range from the magnificent curlew, whose curved bill is so long it has special nerves at the tip to feel its wriggling prey, through godwits with straight bills, oyster catchers with thicker bills, perfect to catch not oysters but mussels, redshank with long scarlet legs, right down to knot and dunlin with tiny bills, with which they probe the mud like little sewing machines, up, down, up, down. Even without binoculars, you can see a great range of birds if you time your visit to just before high tide when the small amount of feeding area still uncovered by the seawater means that they are crowded together close to where you are watching. With binoculars and a low winter sun behind you, you are in heaven as you feast your eyes on all the different colours of legs, bills, breast feathers and eyes. And when they are spooked and a whole flock takes off, the different colours as they wheel in the air is magical. There are similar mud flats in Strangford Lough in County Down and in Lough Swilly in County Donegal – all wonderful places to visit in winter.

But then, birds are easy to see when you know where to go, as these big flocks find safety in numbers. What if it is mammals you are interested in? Irish mammals have every reason to fear humans and keep well away from them. The captivating pictures photographers get are taken after long hours spent waiting in hides. The less dedicated observer must go looking for mammals when they are so distracted that they can't be bothered being cautious and afraid. What could encourage such a state? What indeed but the usual – a great rush of testosterone – that overcomes normal bashfulness. Deer do not pair off in couples – a male and a female, with their adorable little fawn. That's only in the Bambi stories. There is a harem set-up with deer, where one dominant male beats off all the other males and has all the females for himself. Achieving this pole position is not easy and the stags fight it out during the rutting season. If you want to see deer without having to crawl on your belly through midge-filled uplands, then go looking during the rutting season when they have more important things on their minds than nosey humans.

The rutting season for Red deer is in full swing by late September and may continue until November. During this time the stags, who have spent the summer in male bachelor company, join the female group of hinds. They bellow and roar to show what fine fellows they are and to keep other potential male suitors away. If the roars and posturing don't work, they will lock horns and wrestle and try to push the upstart backwards. These battles can last up to thirty minutes at a time, and the sound of the antlers clashing can be heard from quite a distance away. This is the time to visit Killarney National Park in County Kerry, where our only true native Red deer herd lives. There are also Red deer in Glenveagh National Park in County Donegal (which have Scottish genes because of introductions over the years), and the rut can also be seen and heard here from the road up through the glen.

We have two other species of deer in Ireland, both introduced – Sika deer and Fallow deer. The males of both these species, stags in the case of Sika deer and bucks in the case of Fallow deer, also conduct noisy rutting behaviour in autumn to impress the ladies and keep away potential rivals. Rutting Sika stags can be seen and heard in the Glendalough valley in County Wicklow from late August to early December, while you can observe these same shenanigans in Fallow bucks if you visit the Phoenix Park in Dublin or Doneraile Park in County Cork or Portumna Forest Park in County Galway during October.

Giving birth is a full-time occupation as well and, as you can imagine, mothers are concentrating on their own affairs at this time. Seals are mammals, and they have to come ashore to give birth. In the case of the Grey, seal births take place from mid-September to mid-November, and large groups of pregnant females assemble at the breeding grounds from August onwards. The newborn pups are fed with the very rich mother's milk and do not take to the water until they are weaned, which may be up to a month later. Before they all go off back to sea again for another year the males arrive for the mating season.

Needless to remark, there are the usual testosterone-fuelled fights among the males, with each male trying to mate with as many females as he can – no cosy couples in the Grey seal world either. Females leave soon after mating, but the males stick around to the very end. While it is very important not to disturb females when feeding as this can cause pups to be abandoned, seal colonies can be seen from a distance and binoculars used to get good views.

There are Grey seal colonies in Wexford and on offshore islands, such as the Innishkea Islands in County Mayo and Great Blasket in County Kerry. Common seals, which are less abundant than Grey seals, give birth in June and July, and our largest populations occur in Strangford Lough and Dundrum Bay in County Down.

If your interest is in flowers and plants, rather than the world of animals, then at least you are not looking for a moving target. Different habitats put on a display at different times of year, depending on environmental requirements. Deciduous woodlands are great places to look for spring flowers before the canopy of leaves closes in late spring and blocks out the light. The largest planted beech forest in Western Europe is Mullaghmeen in County Westmeath. At 258 metres (846 feet), it is the lowest county top in Ireland and easily accessed. A spring walk here is exceptionally beautiful.

If it is world-class wildflowers you want to see, then it has to be the Burren in north County Clare in the month of May. This area of limestone pavement boasts two different situations side by side. Flowers that are native to the Mediterranean regions of Spain and Portugal, such as the Maidenhair fern, the Large-flowered butterwort and countless species of orchids grow cheek by jowl with flowers from Arctic Alpine regions, such as the piercingly blue Gentian, and the creamy Mountain avens, and the recently re-discovered Arctic sandwort, a small white-flowered cushion plant. Cuckoos abound here still and the amazingly named 'legless lizard', which looks like a snake but is actually a slowworm, occurs here too – an introduction in the last century.

We have long stretches of sand dunes where the tidal currents are slowed and sand is deposited ashore. Much of the coast of counties Wicklow and Wexford is sandy, with the shoreline backed by rolling dunes. On the west coast the sand dunes form on the lee side of headlands, where the slowing currents deposit the sand. Golf clubs have laid out most of the large areas of fixed dune as links golf courses, which may be fine for golfers but present a modified habitat for the former ground-nesting birds, such as skylarks. But further north, from Galway right up to Donegal, the westerly gales are so strong that often instead of a series of dune ridges we have a type of habit called machair. Machair comes from the Irish word *macaire* – which means a plain, and this is in effect a flat sand dune system. These are mosaics of dunes, grasslands and wetlands and grazing by sheep and cattle keeps them this way. They are considered to be priority habitat under our European Habitats Directive and are important for rare breeding species such as Red-necked phalarope, corncrake, choughs and dunlin. You can visit them on the Mullet Peninsula in County Mayo, but be warned, it is always windy there.

If you want to see the most exotic beach in Ireland, then you need to go to County Galway, where you will find a place where there is no sand, just what appears to be tiny broken pieces of coral all over instead. This is not actually coral at all but parts of two red algae that grow in the sea water nearby and absorb lime from the sea water, which they excrete as calcium carbonate. When the algae are washed up and break down, the result is a white beach, formed from millions of fragments of these algae. You can see this at Mannin Bay, beside Clifden. Well worth a visit.

Ireland has wonderful places still for wildlife. Faraway hills are not greener and as travelling to such faraway hills becomes less and less sustainable – if it ever was – it is more and more vital that we appreciate what we have at home.

10

BACTERIA

We couldn't live without them

Bacteria should get themselves new public relations staff. We think of them in a very bad light – causing disease, spoiling food, making life very difficult for everyone. If we could wave a magic wand and make them all extinct, I am sure some people would think that that would be a very good idea indeed. Well, it wouldn't. If we made bacteria extinct, then it wouldn't be long until we were in dire straits ourselves.

Allow me to explain. Bacteria are single-celled living things. They do not make their own food as green plant cells do. They get their nourishment to reproduce and multiply in other ways. Some of them feed on no-longer-living cells that contain food they can use. In the process of acquiring this energy they break down these cells. So, they are decomposers in other words – really useful engines.

They are vital in breaking down what is politely known as organic waste – such as sewage, garden plants in the compost heap, autumn leaves, corpses and the like. Generally, as living things, they need oxygen to do this. They are very good at getting first dibs on any available oxygen and if that is in short supply then the bacteria act quickly. There is 20% oxygen in the air, which is enough for everyone, but water has about 5% dissolved oxygen available for all the different creatures that live there and we are talking about really clean water here. Increase the amount of organic material in water – such as a sewage discharge, or illegal dumping of waste food, or milk, or silage run-off – and the few bacteria that are always present cannot believe their luck. It might seem that they must contact their bacterial cousins on their mobile phones and invite them round to share the feast, so quickly do the numbers of bacteria increase. What actually happens is that the existing bacteria increase exponentially, reproducing every twenty minutes. In the space of just seven hours, they can go from one to a million.

They proceed to take all the oxygen out of the water as they go about breaking down the organic waste – leaving nothing for the fish, or the dragonflies, or the mayflies, or any of the law-abiding fauna for which the fresh water was their natural habitat. Bacteria have been on this Earth for 3.5 billion years, so they know a thing or two about surviving. Taking their turn with available oxygen supply or sharing it is utterly foreign to them.

But imagine a world where organic waste never broke down. All the nutrients would be tied up and not recycled to allow more plants to grow and start the cycle again. Not good. Now it is true that bacteria are not the only decomposers – fungi give a hand in the process too – but sewage breakdown is mainly a job for bacteria.

So what are you doing pouring bleaches down the toilet – 'to kill all known germs dead'? This is complete madness if you have a septic tank, and there are half a million such tanks in Ireland. How do you think the organic waste you and your household have produced will break down in the absence of bacteria to do the job? The answer is that it is not broken down. It is supposed to be broken down in the soakaway as it leaves your septic tank – one flush at a time. This cannot happen if there are no bacteria around. Instead it will soak down through the soil into the groundwater, making the water not fit to drink. As the people who lived around Lough Corrib in houses with septic tanks discovered, when they tried to go back to their original wells when the mains supply from Terryland in Galway was down for six months because of contaminated water. Bottled water sales were never so high.

If you live in an area where your sewage goes off down the mains to the water treatment plant, it is a different story. In Dublin, for example, the waste water is treated at Pigeonhouse Road. Long ago, it was just discharged, untreated, into Dublin Bay, where it went to feed the shellfish. The cockles and mussels there were second to none due to the plentiful supply of organic material they consumed from filtering the seawater. But they also accumulated bacteria from sewage. If there had been a particularly nasty epidemic of, say, typhoid fever in the city, the cockles and mussels would be a further source of catching the disease. No wonder Molly Malone died of a fever. It was the end of most of the O'Connor family too – the mother and four of their five children, as immortalised in James Joyce's *Ulysses* – who ate mussels collected near a sewage outfall at Seapoint in Dublin Bay. EU regulations mean that uncleaned discharges are no longer allowed from the waste water plant and Dublin Bay has now become so clean that Dollymount Beach on Bull Island qualifies some years for a Blue Flag, so excellent has it become.

We would be in a bad way too without the bacteria that inhabit our large intestine and help us digest our food. Complex carbohydrates are broken down by the bacteria that live there. We only miss them when we have killed all our resident bacteria with strong antibiotics. These carbohydrate-breaking-down bacteria really shine in the stomachs of cattle and other ruminants that feed on grass. Humans could never digest grass – too much cellulose – as the poor starving wretches who died with grass stains on their mouths during the Great Famine despairingly found out. The bacteria in the rumen of the grass-eating cows break down the complex cellulose molecules, allowing the cattle to absorb the goodness from grass and thrive. These specialised bacteria ferment the grass and break it down. Very little oxygen is required for this, and the fermenting process releases such by-products as carbon dioxide and methane – two gases that the cow gets rid of by belching them out into the air. Much more methane than carbon dioxide is produced. A dairy cow can produce between 250 and 500 litres of methane and six litres of carbon dioxide per day. I am sure the cow feels a lot better having belched all this out, but with 6.5 million cattle in the country these emissions of greenhouse gases add up.

There are of course lots of bacteria that get their nourishment from living cells and these are the ones that cause us problems and deserve their bad reputation. When these ones enter our bodies they quickly multiply if conditions are right and they can cause us to have very nasty diseases indeed. Cholera, typhoid, tetanus, and tuberculosis among others are diseases caused by different types of bacteria. The Black Death or bubonic plague that killed a third of the population of Europe in the mid-fourteenth century was caused by a bacterium spread by fleas, as was typhus – the famine fever that killed as many people as hunger did in the 1840s in Ireland.

But they didn't kill everyone. The human body can fight back by producing antibodies to the bacteria in question. When the disease is finally routed these antibodies can remain in the white blood cells and clobber any further infections from those same bacteria if they should try to infect at a later stage. Immunity in other words. Vaccines work in the same way. The disease-causing organism is de-activated so that it won't cause disease but will still stimulate the white blood cells to develop antibodies and then – after a great deal of testing to make sure it is absolutely safe – is injected into us in order to make us immune to the disease in question. The BCG vaccine broke the stranglehold of tuberculosis that had devastated human lives right up to the 1950s.

So if you are immune, the infecting bacteria cannot take hold. But if you are not, then what medicine can help? Something that will kill the bacteria obviously – in other words antibiotics, which were first discovered by Alexander

Fleming in 1928. He noticed that his bacterial colonies growing in agar plates in his laboratory were killed if his petri dishes got contaminated with fungal mould. Instead of just binning them and being more careful the next time, he examined them to see how it actually happened. He discovered that something seeped from the fungal mould through the agar jelly to the bacterial colonies and killed them. Because it came from a fungal mould called *Penicillium* he called this substance penicillin.

It proved very useful in killing all sorts of bacteria and, in due course, medicine containing penicillin was used to kill bacteria causing disease in humans. It killed living things – bacteria – so these medicines were called antibiotics (from 'anti' + Greek *biotikos*, meaning 'fit for life'). But bacteria haven't been around for 3.5 billion years for nothing and many of them rose to the challenge. As they have such a quick reproductive rate, any mutation in the bacteria that caused it to be resistant to the antibiotic could quickly give rise to a new strain. It is an arms race between the mutating bacteria and the scientists who are producing new sorts of antibiotics all the time. Our own attitude to antibiotics does not help matters – in fact it is possibly leading to catastrophe unless we cop ourselves on.

This is why. You have a terrible sore throat and a chest infection and go to the doctor. He diagnoses correctly that this is a bacterial infection and prescribes antibiotics for you. You are told to complete the dose of tablets or medicine which may be for a period of five to seven days depending on the type. You take these for three days and are feeling much better indeed. After another day your large intestine tells you that it is finding it difficult to digest the complex starches that are normally digested there – the antibiotics are not confining their attentions to just the bacteria in your throat and chest, those in your gut are getting a wallop too. And many of us give up the antibiotics at this stage and don't finish the prescribed dose. What have we done?

Look at the big picture. Your sore throat is caused by a bacterium called *Streptococcus*. There are lots of them there – a whole range of individuals. The first two days of antibiotics clobbers most of them and you begin to feel better. But there are always the hard chaws that don't respond to the first few goes. You have to keep taking the tablets until every last one of them is killed. If you stop too soon those hard chaws are still around – cheering no doubt because they weren't killed. They bide their time, increase and multiply and, in the fullness of time, become resistant to that antibiotic. By not completing our dose we are allowing the most resistant ones survive. (Our guts will be grand, even after seven days; we can take probiotics and restore the good bacteria there.) But if *Streptococcus* bacteria that are resistant to antibiotics are being created, and we have run out of different antibiotic treatments, well then we could end up with a bacterium that cannot be contained. This is exactly what MRSA (methicillin-resistant Staphylococcus aureus) is — a 'superbug' that not even the excellent antibiotic methicillin can kill. Therefore, the plea is: FINISH THE DOSE.

And, indeed, stop demanding antibiotics for diseases not caused by bacteria. They won't work. What else causes diseases? Well viruses do. A whole new chapter is required to examine this.

11

PANDEMICS

**Or why you cannot get Covid-19 or any
other disease from telephone masts**

Many human diseases are caused by germs. When there is a widespread occurrence of a disease in a community at a particular time it is called an epidemic. When it spreads to many countries, or worldwide, it becomes a pandemic. Throughout the history of the human race, we have had many such widespread diseases. In earlier times it was not known what caused these diseases or how they were transmitted, and they were often interpreted as being sent by an avenging god as punishment for sin. Many of these diseases were caused and spread by bacteria – such as the Plague of Justinian in the sixth century and the Black Death in the fourteenth century – both of which were epidemics of the bubonic plague. Not that anyone knew about bacteria in those days.

But some diseases are not caused by bacteria but by viruses, although that was not apparent originally. Smallpox is caused by a virus and it has been around for a very long time indeed. Evidence of it has been noted on Egyptian mummies that are three thousand years old. It was widespread throughout the known world and greatly feared – both because of the high death rate and also because survivors were commonly left with ravaged faces from the permanent scars caused by the disease pustules.

It was Edward Jenner, an English doctor, who, in 1796, noticed that milkmaids who contracted a very mild disease called cowpox never contracted smallpox, although it might be widespread in the area. He concluded that something in having the cowpox disease caused immunity to the very serious smallpox. He experimented by injecting material from a cowpox pustule into the arm of the nine-year-old son of his gardener and, later on that year, he repeatedly exposed the lad to smallpox, but the child never caught it. We don't know what the gardener, or indeed the child's mother, thought of such highly questionable 'experiments', but Jenner was on to something. The Latin word for a cow is *vacca* and he called his method vaccination – injecting a disease of cows into humans made them resistant to smallpox. The word stuck.

Bacteria were first seen in 1670 when advances in making glass lenses allowed them to be discerned in water samples examined under strong magnification. No connection was made between the bacteria and diseases until the mid-1800s because everyone of course knew in those days that diseases were caused by miasmas – poisonous vapours in the air, particularly the night air – which were alleged to come from decaying vegetation. In fact, the widespread disease malaria gets its name from bad air. Although malaria is spread by bites from infected mosquitos, it is common in marshy areas as mosquitoes need still, fetid water in which to breed. But it is not miasmas that spread malaria or any other disease.

It took the work of Robert Koch, in the mid-nineteenth century, to establish what was called the germ theory of diseases. He proved that cholera and tuberculosis were caused by bacteria that could be seen under strong light magnification. And subsequently many other diseases such as syphilis and dysentery, to mention but a few, were discovered to be caused by bacteria as well. But there were many diseases that were not caused by bacteria – polio, rabies, measles and mumps and the so-called Spanish flu that killed up to 50 million people worldwide in 1918/1919. What caused these, very contagious, diseases?

Viruses were not discovered to exist until the end of the nineteenth century. The name virus was given to the cause of a disease of tobacco – tobacco mosaic disease. It described something so small as to be able to pass through filters fine enough to keep back bacteria and yet infect other healthy plants. In 1901, yellow fever was the first human disease to be attributed to a virus. It was to take until 1931, when electron microscopes were invented, before there was technology that allowed us to see exactly what a virus was, but it was to be 1955 before images of the full structure of a virus – again the tobacco mosaic virus – were first produced by Rosalind Franklin.

What was this thing – a virus? It wasn't a living cell. Living cells, even the simplest of them such as a bacterial cell, had cell walls and genetic material – DNA – inside on a single chromosome. Viruses – fifty times smaller than a bacterial cell – seemed to be just genetic material – DNA – surrounded by an outer coat or membrane. How they work, it was discovered, is that when they enter the host – a living thing – the outer coat of the virus sticks to the cell wall of a chosen cell in the host. The virus penetrates this cell wall, allowing the DNA in, and this then takes over the host's cell's DNA, causing it to make lots and lots of virus. The cell then bursts, and all the new viruses are released to go on and attack new cells. If the living host is not able to put up any resistance, it is easy to see how they are quickly overwhelmed.

There was a disease called 'the Sweat or Sweats' in Tudor times in England. Apparently the foreign soldiers of Henry VII brought the disease to London when they entered the city in triumph after defeating Richard III at the Battle of Bosworth. Henry VII's son, Henry VIII, lived in dread of it and left London every summer so as to escape it. This disease killed quickly; people who were well in the morning could be dead by the following one if they caught it. This disease – which seems to no longer exist – was thought to have been caused by a hantavirus, to which there was little resistance.

Lots of common diseases were found to be caused by viruses, but everybody didn't die from them, so the body must have been able to fight back. And yes, indeed it can do, by producing antibodies that latch on to the virus and stop it being able to stick on to the cell wall of the host and get its DNA inside. In the best scenarios there are always some of these antibodies around in the host to attack the same virus if it ever shows up again. This is what Edward Jenner had inadvertently stumbled on. The antibodies formed by the body to fight cowpox virus were able to stop the more deadly smallpox virus too – luckily for the gardener's son.

We now know there are different sorts of viruses. Some have a helical or rod shape, like the Tobacco mosaic virus. Some have a spherical shape, and some specialise in infecting bacterial cells. Many of them have DNA as their nucleic acid. But there are some – called retroviruses – that have RNA instead. This is a different type of nucleic acid that is converted into DNA when the virus infects the host cell. This then is able to make the host DNA make new retroviruses with RNA inside. A step smarter. The virus that causes AIDS is a retrovirus called HIV. So you can see why antibiotics are no good against diseases caused by viruses. Antibiotics attack bacteria. There are no bacteria implicated in straightforward viral diseases. So, the advice is to stop expecting your doctor to prescribe antibiotics for the virus that causes the common cold. Those antibiotics will only annoy the good bacteria in your guts and maybe make other not so good bacteria immune, so that superbugs are created.

In late 2019 a virus, unknown before, affected the lungs of people in the Wuhan province of China, killing many and spreading rapidly in airborne droplets produced by sneezing and coughing. Its breed, seed and generation were rapidly studied, and it was discovered to be a spherical virus whose nucleic acid content was RNA. The sphere had projections that matched with human lung cells, and when they came in contact with lung cells they hooked on, thus allowing the RNA to enter. The invaded cell then made lots more viruses that went on to infect and, in the worst cases, kill the person who was infected. Because the coating with the projections looked like a crown in the images, it was called a corona virus, and because the disease caused by it began in 2019, the disease was called Covid-19.

Where had this virus come from? Why was it not known about before? Needless to say, all sorts of conspiracy theories abounded. But where do viruses come from anyway? Any virus? What do we know about their origin? They are not living things, as such, as they cannot reproduce by themselves. It is thought that they are pieces of nucleic acid that once were part of living cells but escaped, as it were, and evolved an ability to acquire a protective cover around their nucleic acid core. There have always been viruses specific to certain living hosts. They need to invade the cells of these hosts in order to replicate themselves. The protective coat is everything. If this cannot lock on to the host cell wall, then the viruses cannot get their nucleic acid into the cell and take over its workings. Wild animal species have their own viruses which are specific to them.

It seems that if we stress wild animals enough, destroy the habitats where they live or capture them for wildlife trade, it is possible to cause some of their viruses to change slightly so that they can lock on to our cells. The so-called Spanish flu pandemic of 1918 is thought to have leaped from poultry to WWI soldiers who were in close proximity. Bird flu, which affected humans first in 1997, was a disease of wild fowl that spread to domestic poultry and then jumped to humans who were in close contact with the poultry in crowded stressful conditions. SARS was a disease that was identified in 2003. It was caused by a corona virus that moved from wild animals, possibly bats to civet cats and then to humans. Covid-19 is thought to have spread from bats to pangolins to humans.

Global movement of people spread it to every continent with huge death tolls. People who have recovered from it have antibodies to the virus but how many antibodies they actually have, how long-lasting these are, and thus the protection they offer, is not clear. Vaccines from de-activated virus or from replicating the virus coating with no nucleic acid involved take time to make and test. Vaccines for the usual winter flu only confer protection for a year and people have to get the flu vaccine every year. How long will a vaccine against Covid-19 give protection? The world's response in 2020 while waiting for such treatments was, in fact, the same as the medieval responses to the Black Death. Stay away from those who have it or might give it to you.

And where do the telephone masts come in?

Diseases do not emanate from 3G, 4G or even 5G masts. The masts work using electromagnetic radio waves. There is of course a whole spectrum of electromagnetic radiation. It ranges from those with the shortest wavelengths and the most energy – such as Gamma Rays, X rays and UV rays all of which contain enough energy to damage living cells – through the visible light spectrum, right down to those with the least energy and the longest wavelength – the radio waves. These are the ones used for computer networks, radio transmissions and such. Without them we would have no mobile phones, no computers no radios. New masts are designed to improve the transmission network.

Masts have absolutely nothing to do with disease transmission. Radio waves cannot affect our living cells in any way or make us more susceptible to diseases of any sort. While individuals may not understand this fact or how the world of masts works, the fact that they then feel entitled to go on to attack masts and seek to deprive whole areas of vital communication networks beggars belief.

We have the expression 'a little learning is a dangerous thing', but what is even worse is no learning at all or misguided notions that can be catastrophic.

12

MODERN TERMS

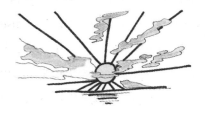

**That we are all supposed to know,
but weren't paying attention at the time**

As recent events in 2020 have made abundantly clear, we all need to know how the world works – not just the scientists. The narrative has always been there, but unless you were following it all the time – and really who was? – many of us have our own understanding, or not, of the terms being used to describe the world we live in.

These terms have crept up on us. They are all top stories in the media at a given time. Then a new crisis occurs and we never hear how the earlier story turned out. Remember the ozone layer and the holes in it? Are they still there? What is coming through them? Is it all fixed?

And are global warming and climate change the same thing? Are they two different things? And what is really making it happen? And hasn't it always been happening since geological times? Why are humans being blamed now?

Does burning rubbish in our back gardens make the holes in the ozone layer and allow in more heat from the sun thus causing the world to warm up – as one former minister for the environment postulated in the not so very dim and distant past?

Understanding how these things happen doesn't have to be long-winded or complicated. This is how the world works.

The sun, our nearest star, has a big role to play. The radiation that comes from the sun is crucial to life on Earth. All this radiation contains energy and it must pass through the atmosphere to get to Earth. Our atmosphere surrounds Earth and forms a layer, whereas the moon, for instance, has very little. We can actually see some of this radiation – the visible spectrum of light. It looks white, but as any rainbow shows us, it is really a range of colours from violet to red. At least that is what our eyes tell us. Other creatures we share the planet with see things differently. Bees' eyes cannot see red for example, but then they can see beyond the violet into the ultraviolet. Neither can bulls see red – so like a red rag to a bull is not so. A yellow rag or a green rag would have the same effect; it is the movement of the rag that will enrage the bull if he is going to get mad, not the colour. To be exact, it is the reflection of light from objects that our eyes see. Green leaves absorb all light except green – this is reflected – and so they look green to us. If all the light is absorbed and none reflected, then the object looks black, and similarly if it is all reflected, things look white.

So far so good.

Light from the sun travels to us in waves as indeed all the radiation from it does. The wavelength of blue light is shorter than that of red light. If we go past the visible spectrum into the ultraviolet region, the wavelengths get shorter and contain more energy. This is not necessarily a good thing for the human race – rays full of energy can damage living cells. Just as well the Earth has some protection from these very energetic rays, or we wouldn't be here at all. The protection is the 18km of our atmosphere closest to the Earth. The rays from the sun have to travel through this in order to get to Earth. So what is this shield, the atmosphere, made up of? Mainly nitrogen (78%) and oxygen (21%) with other bits and pieces making up the remaining 1%. When it comes to protecting us from the effects of very energetic ultraviolet rays, it is oxygen that has the starring role.

The oxygen molecules in the atmosphere consist of two oxygen atoms bonded together – O_2. Well, it does in the main. But 18km out in space conditions are different. The atmosphere is much thinner, pressure is much lower, and it is very, very cold. The poor oxygen is terrified and goes around in gangs of three holding hands as it were – O_3 – and in this form it is called ozone. This ozone is created by the UV rays coming in from the sun. They split the outside layer of O_2 molecules and the individual oxygen atoms bond with the other O_2 molecules to form the ozone. There is an ozone layer all around the Earth at this height and the energetic ultraviolet rays cannot get through this but are reflected back into space.

So we are grand, or at least we were until we discovered a way to damage this ozone layer, break up the gangs of three, and turn them back into O_2, thus making holes in the good ozone layer, allowing ultraviolet rays get through and cause skin cancers.

Of course, we didn't know we were doing this. It was only in 1985, when a scientific paper appeared in the science journal *Nature*, that this was announced to the world. And how were we damaging it? Well, our chemists had invented a gas called chlorofluorocarbon (CFC) in 1928, which was a much safer gas to use in the making of refrigerators than previous toxic gases such as ammonia and sulphur dioxide had been. In fact, these stable CFCs were so useful that they were widely used in aerosols, air conditioners and expanded polystyrenes as well. At its peak more than a million metric tonnes were produced annually.

But while these CFCs were inert in the lower atmosphere, in the upper atmosphere it was not so. The chlorine part got released here and attacked the ozone, turning it back into O_2, through which the UV rays could pass. And it happened fastest in the extreme cold and dark of winter over the Antarctic where this great hole was discovered. The ultraviolet rays were able to get through and give skin cancers to living things exposed to them. Nobody really lived in the Antarctic, so was that really a problem for us?

But it was soon discovered to be happening over the Arctic as well. And the UV rays were getting though and were posing risks of skin cancer to the populations of northern Europe and North America, among others. This wouldn't do at all. So, at a convention in Montreal in 1987, a protocol was drawn up to stop the production of CFC. And thus, it happened. Fridge coolants no longer have chlorine in them and we are all saved. However, fridges are long-lasting pieces of equipment and many of the machines made using CFC as coolant are still around and have to be disposed of carefully. This is another good reason not to dump them in the nearest bog when you are upgrading your kitchen.

The holes in the ozone are getting smaller. Mind you, we had made so much CFC up to 1987 that the holes continued to get larger until the year 2000. But is has been slowly decreasing year on year since then and we should be back to 1980 levels by 2070. If we are around by then that is.

Global warming however is a different kettle of fish, as it were.

Back to the sun and the energy coming from it towards Earth. This time we are looking at the rays that are longer than the red rays of the visible spectrum. These are the infrared rays and are what heat up our planet. These come through the atmosphere all the way to Earth and warm us up. They bounce off the Earth and go back out into space – just as a ball might bounce off a wall and come back if you threw it. And just as hitting the wall stops the ball, changes its direction of travel and causes it to come back slower, so does hitting the Earth cause the infrared rays to change direction and be reflected back to space through the atmosphere at a slower speed than they entered it. When these rays are slowed down, they have a longer wavelength than before. When they reach our atmosphere on their way in from the sun, they all can pass through to Earth, but on the way out again, with their now longer wavelength, the situation is different. They can no longer pass freely.

Even though 99% of the atmosphere is composed of

nitrogen and oxygen, the components of the other 1% are vital for life on Earth too. One of these other components is carbon dioxide. In the great scheme of things there is very little of it actually – at the moment it is a mere 0.0415%. So little in fact that we usually refer to it in parts per million – ppm. (0.0415% is 415 ppm.)

One of the reasons why carbon dioxide is vital for life on Earth is because the slowed-down heat waves cannot pass back out through it. It there was none in the atmosphere, all the heat rays would bounce back into space and we would be very cold indeed here. If there was so much that none of the slowed-down heat waves could get back out again, the planet would boil. But like Baby Bear's porridge in the Goldilocks fairy tale it is just right, and enough heat is retained to allow life to have evolved on Earth over the last 5 billion years. In fact, the term Goldilocks effect has crept into the scientific literature on the matter – proving that scientists must have been young once.

So, what's the problem then? What's all the hullabaloo about? Well, the thing is that the amount of carbon dioxide in the atmosphere has increased very quickly and is now preventing the escape of more and more reflected heat rays and causing the temperature of our planet to rise. Reputable and verifiable scientific measurements report that in 1800 the amount was 280 ppm and now it is 415 ppm – a rise of 48% in just over two hundred years. Now, we did have that amount of carbon dioxide in the atmosphere before – analysis of ancient air bubbles trapped deep in Antarctic ice gives the composition of the air at the time these bubbles were trapped. And so it was 3 million years ago, during the Pliocene Epoch when human life had not evolved yet on Earth, when carbon dioxide levels in the atmosphere were last so high. For the past one million years – right up to 1800, the amount of carbon dioxide remained remarkably steady, never rising above 300 ppm. It is our relatively sudden rise to 415 ppm in the short period of 220 years that is giving rise to such concerns with a lot more heat from the sun being retained.

Why? Why has the amount of carbon dioxide increased so much in such a short period of time? And is it really the fault of humans?

The first thing to consider is where does the carbon dioxide in the atmosphere come from? There is a carbon cycle, where plants take in carbon as they grow; carbon then moves to the animals that eat the plants; they are all broken down by the decomposers – fungi and bacteria – when they die; and the carbon dioxide is released back again into the atmosphere. A stable cycle you might think, and so it was for the last million years or so. However, not all the plants that died were broken down by the fungi and bacteria. These are living things, too, and if conditions are not right for them, they can't do their job. So plants and indeed animals that ended up in frozen wastes too cold to allow decomposers, or in wet acid conditions with no oxygen – such as we see in bogs to this day – or other unsuitable circumstances, well then, they never broke down and the carbon they contained was never released back into the atmosphere. Over the hundreds of millions of years of life on Earth, these became compressed into coal deposits or oil or natural gas, depending on environmental circumstances (what we call fossil fuels). The normal vegetation of Earth – mainly forests, whether tropical rainforests, temperate forests or cold coniferous forest, depending on climate – kept the carbon cycle going and the carbon dioxide levels in the atmosphere in a more or less steady state over the past million years.

There is no getting away from the fact that it is the behaviour of humankind that has caused the huge increase in atmospheric carbon dioxide. What have we been doing? Two things actually. First of all, we have been removing the busy carbon-capturing forests of the Earth. Secondly, we have been releasing the carbon that was captured millions of years ago when those deposits of coal, gas and oil were forests. We are putting much more into the atmosphere and slowing up the mechanism by which it is taken out. And doing it over a very short period of time – as 220 years is, in the Earth's scale of counting. Elementary, my dear Watson.

We well know what the story is, but unlike the hole in the ozone layer, we haven't got around to fixing it. Not yet, anyway. Not until it finally dawns on us – hopefully before the horse bolts – that how we live as a species is changing the very world we live in.

13

GLOBAL WARMING/ CLIMATE CHANGE

**Is it getting hotter or colder
or wetter or drier or what?**

Global warming is happening because there is more carbon dioxide in the atmosphere now than there used to be. But is that not a good thing for Ireland? After all we could do with a bit more heat in the summer. At the moment if we get twenty-five degrees Celsius for more than two days on the trot, we think we are having a heatwave, so unused are we to nice hot weather.

Well, it may surprise you to learn that we already are far hotter than we should be, relative to our latitude, and this has been the case for the last ten thousand years or so since the end of the last Ice Age. Allow me to explain. Suppose you could go directly in a straight line across the Atlantic Ocean to North America, where would you end up? If you went directly across the sea from Galway, you would come ashore in Labrador, just to the north of Newfoundland, which is the same latitude as Galway. And the climate there is much colder, particularly in winter. In Galway there is very rarely any length of a spell when the temperature drops below zero Celsius in winter, whereas in Newfoundland and that part of Labrador, temperature is below freezing from November to March, while in summer it rarely reaches twenty degrees Celsius.

What makes Ireland punch above its weight and also keeps average temperatures down in Newfoundland are the currents in the Atlantic Ocean, off both their coasts. Ireland, as we were all told at school, is washed by the North Atlantic Drift, also called the Gulf Stream. This is a warm ocean current that flows northwards along the western coast of Europe. As the southwesterly prevailing winds move over it, they are warmed up and so bring temperate winter conditions to our shores. As far north as Narvik, which is well within the Arctic Circle, the harbours are ice-free in winter thanks to the warming effect of this ocean current.

But what goes up must come down. The water eventually cools, sinks and returns south again as the current moves in an anti-clockwise direction. The cool returning current passes along the southeast coast of Greenland and meets the cold Labrador current flowing south along the western coast of Greenland. This union of currents then passes the coasts of Labrador and Newfoundland, making their winters longer and colder than those of equivalent latitudes in Western Europe.

But as more and more heat is trapped by the increasing amounts of carbon dioxide in the atmosphere, the permanent ice in the Arctic ice cap and in the Greenland ice sheet is being reduced. This is having an impact on the North Atlantic Current. The Greenland ice sheet is made from fresh water. With rising temperatures, great glaciers slide down to the coast more rapidly than ever and great chunks of freshwater ice melt and go into the ocean. Think of the effect this is having on the North Atlantic Drift.

Hot water is lighter than cold water, so as the current flows northwest through the Atlantic from the Gulf of Mexico, where it originated, the warm part of the current is at the surface. Thus, it warms the winds that pass over it, raising the temperatures of our shores in winter. As the current moves further north, it continues to cool and eventually as the cooler water sinks, the current swings around and moves in a westward direction along the south-east coast of Greenland. As it goes along, it is now receiving huge quantities of cold fresh water from both the melting glaciers and from the run-off from the warming ice surface in Greenland. Fresh water is lighter than salt water, so this cold fresh water is now sitting on top of the cold salt water and the effect of this is to diminish the North Atlantic Drift. The water doesn't sink because it is fresh water on top of salt water and so the force driving the current is being reduced. It is reckoned by scientists that this North Atlantic Drift has slowed by at least 15% over the past sixty years.

The more it slows down, the less warm water that moves north along our west coast. When it stops altogether, Ireland will experience colder winters than at present. So, while global warming causes climate change, it may not necessarily mean that our particular climate will get warmer when this eventually happens. But that is not even the whole story.

The ice caps over both the Arctic and the Antarctic are being affected as the world warms up. Satellites have provided images of the extent of these ice caps for the past forty-one years. These caps are always at their smallest size at the end of the summer period, which is September for the Arctic ice cap. When they began measuring it at the end of the 1970s the extent of the Arctic ice cap was over seven million square kilometres. It has been shrinking ever since. In 2012 the lowest extent was recorded – 3.39 million square kilometres – and 2020 was the second lowest at 3.74 million square kilometres. In fact, the ten lowest measurements have all been taken in the last ten years. The Arctic ice cap has halved in area over the past forty-two years and some current models predict that there will be no ice cap at all by the month of September by mid-century. It does increase in size during the colder winter months – at present anyway.

Interestingly the situation in the Antarctic is different – different but not better. The ice cover is at its lowest at the end of summer there, which is at the end of March. This ice sheet lies over the continent of Antarctica rather than floating on the ocean as the Arctic ice cap does. So as the Antarctic ice melts, water enters the ocean from the land and causes sea levels worldwide to rise. They reckon it will have risen by 2.5 metres by the end of this century. That will affect many of our coastal cities around the world, not to speak of entire low-lying countries.

And if all of this weren't bad enough, the shining white ice caps reflect the warm rays of the sun back into space, whereas bare ocean surfaces absorb this heat. So with smaller heat-deflecting ice caps and more heat-absorbing bare ocean, global warming is accelerated – particularly over the Arctic, where it has been occurring at twice the rate of the rest of the world.

The Earth has warmed up on average 1.14^{0}C since records were kept 141 years ago. It may not seem like much, but it took a huge amount of extra heat to increase the temperature of the whole globe by this amount. The scientists tell us that if it goes over 2^{0}C, it will have gone past the point of no return. Already we can see the impact this is having all over the world – and none of it is good.

Remember when we used to say 'a grand soft day' referring to the light rainfall, just drizzle really, that was falling. It made your hair go all curly and frizzy – if that was the kind of hair you had. Nowadays, because warmer air can hold more moisture than cooler air, there is more water vapour up there to fall as rain. So now we have heavier rain, lasting longer, filling our rivers and causing flooding. No more grand soft days and your hair is now plastered on to your head if you are caught out in it without a hat.

Climate change affects different parts of the world in different ways – depending on the circumstances that affect their climates. There were always hurricanes affecting the Caribbean and the southeast coast of North America. These hurricanes form over warm tropical ocean waters. The warm moist air rises up, is replaced by surrounding air that in turn warms up, takes in more water vapour and rises to add to the size of the storm. It cools off eventually as it rises and forms clouds from the water vapour it holds. It spins and grows, fed by the heat from the ocean and moves towards land. It weakens when it makes landfall as it is no longer receiving energy from the warm ocean water. It moves inshore in a great swirling mass and dumps its cargo of rainwater. Hurricanes are increasingly more violent events than they used to be because they are now being formed over ocean waters that are hotter than before. They are called typhoons or tropical cyclones in other parts of the world, but they are getting more violent everywhere they occur, because the ocean waters over which they form are getting warmer.

Africa sees its climate changing in a different way. The sun is directly over the equator at the equinoxes – 21 March and 22 September – and it is after these warmest times that the rainy seasons come. In Kenya and Tanzania, the rains that come from April to the end of May are called the long rains. They have other local names in Ethiopia, Eritrea and Uganda.

Over much of these east African countries these long rains have been declining in recent decades and some years fail entirely in certain areas. Factors governing the variability of the long rains are complicated and have been the subject of much study. But what is not contested is that droughts have become longer and more intense and tend to continue across rainy seasons. This has led to crop failure and even famine. The local people are in no doubt that climate change is causing these conditions.

After the September equinox is another rainy season from the end of October to December. These are known as the short rains and last for about six weeks. These, too, have changed in recent times. On a visit to Eritrea in November 2013 the locals told me that all the rain that would have fallen over a six-week period had fallen in just two weeks, flooding the rivers and rushing down and away to the sea, washing away valuable soil as it ran off the land. The same amount in a normal six-week period would soak down through the soil and replenish the water table so that they could use their wells. This wasn't going to happen in 2013, and it didn't. In these countries where they can get two crops of what they fondly call Irish potatoes (as opposed to sweet potatoes) per year, they were going to get very poor harvests indeed because of the change in the rains.

Australia has a different problem again. Always an arid continent, its indigenous peoples knew how to live in harmony with the environment, having been there undisturbed for more than 40,000 years. The arrival of James Cooke, captain of the *Endeavour* into Botany Bay on 29 April 1770, was to change all that. White people came from the other side of the world to settle in this faraway continent. They had no knowledge of the local flora, fauna and soil. In order to acquire land, they had to agree to certain conditions, including clearing away the native scrub and vegetation. The plants and animals they brought from the far side of the globe were not what would grow happily here. Over the ensuing two centuries the ecology of the continent was radically altered. Now, as the climate heats up, their summers become even warmer and drier. Uncontrollable wildfires sweep through the forests that have been planted and have devastating effects.

So global warming is causing climate change in different ways all over the globe as it impacts on the conditions that affect any particular region. Is the genie out of the bottle? Can we fix it still? There is a simple way, but we have to understand what it is and insist on it happening, and we have to act now and not in some future time when the economy is fixed.

14

MORE BUZZ WORDS EXPLAINED

Carbon sequestration – just a big word for trees?
And what on earth are carbon sinks?

Plant an acorn that weighs 20 grams and come back in twenty years and it will have become a tree weighing a tonne. Where did this stuff in the tree come from? The soil? No, because then there would be a big hole under the tree where that material was. The sun? No, because every time a tree grows does a freckle appear on the face of the sun indicating a piece of it has gone into the tree? It doesn't.

The rain? No. The tree is not just 20 grams plus a tonne of water. So, what's left? The air. Could the tree possibly be made of air? Really? Well, yes actually, it is. A specific part of the air that is – the carbon part of the carbon dioxide that you made the acquaintance of a few chapters back.

Back to the all-important sun again and the rays that come out of it. This time we are focussing on the visible spectrum from violet to red – those colours we see in every rainbow. The raindrops in the atmosphere act as tiny prisms splitting the light into its visible wavelengths as the sun shines through them, thus giving us our rainbow. These light rays contain energy too. This energy can be used by the green leaves of plants and trees. More specifically it is the big green chlorophyll molecules present in every green plant cell that uses light for energy. Put simply, the light energises the chlorophyll and enables it, in the presence of water, to split the CO_2 molecule apart. The carbon part is captured and becomes part of the cells of the plant and the oxygen is released as a gas. This work, done by the chlorophyll in the presence of light, is described as photosynthesis. Without it, there could be no life on Earth as we know it and indeed for some billion years after the formation of planet Earth came about, our only form of life here was bacterial as there was no chlorophyll.

The bigger the plant, the more photosynthesis takes place,

so huge plants like trees take a lot of carbon from the air and store it mainly in the timber. But, of course, why use little words to describe this when there is a perfectly good big word to use instead. So, this taking of carbon from the air and storing it in timber is called carbon sequestration, and evergreen trees in warm countries that grow all the year round store the most carbon. Deciduous trees, such as many of our native trees like oak and ash, only have leaves from May until October so they are off duty for half the year. And evergreen conifers in the northern forests cannot carry out photosynthesis when the temperature is too low. The best rates of carbon sequestration take place therefore in the tropical rainforests – half of which we have already destroyed.

The amount of carbon taken in by a tree is often described as tonnes of carbon or tonnes of carbon dioxide. Are these one and the same. No of course not – why would anything be simple if it can be complicated in any way? It is back to basic arithmetic actually. Carbon is an atom with an atomic weight of 12 on the periodic table that measures such things. On the same table an atom of oxygen has an atomic weight of 16. So, a molecule of CO_2 has an atomic weight of 44 – one carbon atom with a weight of 12 plus two oxygen atoms of 16 each. When a carbon atom is burnt, what the burning process does is cause it to unite with two oxygen molecules and become a molecule of carbon dioxide. That's why you cannot burn things if there is no oxygen present.

Take, for example, a litre of diesel in the fuel tank of the car. This litre contains 720 grams of carbon atoms. When this litre of diesel is burned in the car's engine, each of those carbon atoms unites with two oxygen atoms from the air to make a molecule of carbon dioxide. A weight of 12 becomes one of 44. So, when the litre of diesel is all burned, the 720 grams of carbon present have joined with 1920 grams of oxygen to become 2.64 kilograms of carbon dioxide emitted into the air. Scary isn't it? There is somewhat less carbon in a litre of petrol so when that is burnt in a car engine it emits 2.3 kilograms of carbon dioxide – still an awful lot. It is easy to calculate how much CO_2 is emitted from a full tank of fuel – sixty litres of diesel emits a staggering 158 kilograms of carbon dioxide when burnt.

But trees can take it back out again from the atmosphere by photosynthesis, right? Yes, but. What tree, when, where? A fast-growing, fairly big tree takes in more than a newly planted sapling can, so the figures have been calculated looking at a fifteen-year-old evergreen Sitka spruce tree growing in Irish conditions. This fifteen-year-old evergreen tree can remove 20kg of carbon dioxide from the air in a year and store the carbon as timber. Timber is not entirely made of carbon and nothing else. A tonne of carbon dioxide contains 273kg of carbon, which on average is the amount of carbon in a tonne of timber (allowing for all sorts of variables such as tree species, wetness of timber, etc). This gives us a ballpark figure for calculating whether or not we feel we can emit lots of carbon dioxide into the air and just let the trees take it all out again. Well, they won't – not by a long chalk.

Using up a full tank of diesel – say 60 litres – puts 158 kilograms of carbon dioxide into the atmosphere. That's almost eight fifteen-year-old evergreen trees working hard for a year. Flying to New York and back emits 1.275 tonnes of CO_2 per person. That's almost sixty-four fifteen-year-old evergreen trees and twice that number of more slow-growing native deciduous ones. And indeed, to offset the carbon dioxide emitted on the trip, you should have planted them fifteen years ago.

It takes a Sitka spruce – one of our fastest growing trees in Ireland – forty years to attain maturity. It could then have five tonnes of timber, which means it has accounted for five tonnes of carbon dioxide during its forty years of growth. Our own native oaks which are more slow-growing, take at least eighty years to do the same thing. In 2018 the Environmental Protection Agency (EPA) calculated that Ireland emitted 60.5 million tonnes of carbon dioxide into the atmosphere that year. That's 30,250 hectares tied up for forty years growing fast-growing conifers at a mature size of 400 trees per hectare just to offset one year's carbon dioxide emissions. Given that we only have 7 million hectares of land in the entire country, planting it all with trees would only offset a finite number of years of emissions and of course is not practicable. Apart from the small matter of where we might live and how we might grow crops and food, all of the country is not suited to growing trees anyway. Mind you, we could plant some more than we have already, as we currently have only 11% of the country covered with trees – well below the EU average of 42%. But the fact remains, there is no way we can ever remove all the carbon dioxide we emit into the atmosphere by planting trees and keeping them going for at least forty years.

Would carbon sinks help? What exactly is a carbon sink? Well it is not some new-fangled must-have that has to be acquired the next time you are doing up your kitchen. It is not that kind of sink. It is a store of carbon that has been put beyond use – decommissioned as it were. A carbon sink holds on to carbon in such a way that it cannot escape, unite with oxygen atoms and make a bolt for the atmosphere in the form of heat-holding carbon dioxide gas. There are actually only two ways that carbon can become part of a carbon dioxide molecule. One of these ways we have looked at already. In the burning process it unites with oxygen no bother, whether that process is in the car engine, or in a fireplace or in a runaway forest fire The other way it gets there is through the decomposition process – the part of the carbon cycle carried out by micro-organisms such as bacteria and fungi.

If things are too wet, they won't burn. If they are very acidic and lack any oxygen, the decomposing micro-organisms cannot live there. Which is why bogs, made from sphagnum mosses that grew and captured carbon from the air over the last thousands of years in Ireland, are such splendid carbon sinks. Some of our raised bogs that formed in the great lakes left behind after the last ice age, which ended ten thousand years ago, could have been up to thirty metres deep. That is some volume of carbon stored and put beyond use. Our blanket bogs began to cover our upland areas from five thousand years ago when our climate got colder and wetter and, at its maximum, 11.1% of the whole country was covered. Add this total to the 4.4% covered originally in raised bogs and it is easy to see what an important carbon sink Irish bogs once were.

Coal deposits, and those of oil and natural gas are all carbon sinks too – as long as they stay where they are and are not removed and burned as fuel. Soils can contain lots of carbon from plants that grew there and whose remains are still in the soil – particularly peaty soils. The frozen soils of the tundra regions hold lots of carbon, as it is too cold for any decomposers to break them down.

Our raised and blanket bogs – what's left of them – are a major carbon sink in Ireland. They are our own equatorial rainforests in a way, although to our eyes they may not seem so glamorous or full of biodiversity. Does this matter? Is there a popularity competition among habitats? What is the story on biodiversity anyway?

15

BIODIVERSITY

What did it ever do for us?

The term biodiversity was born biological diversity in 1980 – a scientific term used to describe the fact that there are many different types of living things. It quickly morphed into the term bio-diversity and was being referred to as such in scientific publications from 1988 onwards. Its definition expanded to cover not only the vast variety of living things on Earth, but also the great variety of habitats in which they live, as well as the variation in the genetic make-up of individuals in the same species.

Great counts have been made of everything over the years.

How many mammal species do we have? How many birds? How many creatures with no back bones – the creepy crawlies, although that is not their scientific description, just an evocative one? How many sorts of trees are there? Is it possible to categorise the types of places or habitats where they live? Habitats are controlled by environmental conditions. Some living species are particularly fussy about where they live, others can adapt easily and can be found across a whole range of habitats.

And then there is their genetic variability. If all the individuals of a species have exactly the same genetic make-up, then they are all identical clones. No biodiversity there. If there is more than a certain amount of genetic difference, then we are looking at two species not one. As a result of genetic work towards the end of the twentieth century, it was discovered that 1% was the magic number. Less than that amount of variation between individuals and we are looking at members of the same species. More than that and we are in the presence of two different species that if crossed cannot produce fertile offspring. An end therefore to the hoary jokes such as what do you get if you cross a kangaroo with a sheep and other such gems.

Once we grasped the concept of biodiversity, we were then in a position to actually count everything. We know how many species of everything have been discovered and named. We also produce figures for the number of species that haven't yet been discovered and named – the known unknowns as it were. And once we could count them and produce lists, we quickly began to see that there were fewer numbers of species than there had been several centuries earlier and that the actual populations of these species were declining as well. While we still have tigers, we have fewer of them than we used to have. Fewer whales exist now than when Herman Melville wrote *Moby Dick* in 1851. And if we have fewer species and smaller numbers of individuals, what is the cause?

The reason is quite simply that our wildlife habitats are changing too. Remember the second part of the definition of biodiversity – variety of living spaces. If there is less variety of habitats, then those creatures that can only live in one specific type are in trouble. The one species on Earth that is not in decline is *Homo sapiens* – ourselves. Our population has been growing really fast since 1800 when the world's population was 1 billion. By 1900 it had grown to 1.5 billion and by 1950 had reached 2.5 billion. By the year 2000 the population of this Earth was four times the size it was in 1900. It had reached 6 billion. In the ensuing twenty years another 1.75 billion or so has been added to the total.

All these people have to live somewhere, and they have to eat. The wildlife habitats have been encroached on in order to provide us with food. In many places rainforests have given way to cattle-farming ranches or palm-oil plantations. More fish are caught in the seas than can be replaced naturally. Huge fields of one-crop species attract massive gatherings of the insects that are partial to that crop. The consequence is that it takes a lot of spraying with poisons to keep their numbers down and many 'innocent' creatures are caught in the crossfire.

But we have to live. What does it matter if a few species are no longer around? What good was the dodo anyway? Isn't the country much better off without wolves – the last of which was dispatched by the wolfhounds of John Watson, master of the hounds at Myshall, in 1786 on Mount Leinster in the Blackstairs? We have to drain our wetlands in order to grow crops. After all, people are more important than frogs – aren't they? Weren't all the dinosaurs made extinct and the world is still here?

The dinosaurs were indeed made extinct in what is described as the fifth great extinction which happened 65 million years ago when a meteorite struck Earth, sending massive amounts of debris into the atmosphere. This drastically changed the climate and caused the extinction not only of the dinosaurs but also of 75% of all other living species. And the world didn't recover overnight either. It took millions of years before the species that were not wiped out began to adapt and expand.

Habitats that are no longer good for the species that used to inhabit them are not healthy ones for any species. A habitat with a full complement of plant producers, making food for the herbivores and carnivores that depend on them, is a stable one. The decomposers break down the dead plants and animals, and recycle the nutrients taken in by them to enable the plants to continue growing.

Our richest habitats on Earth – the tropical rainforests – have a huge variety of plant and animal species and have been stable ecosystems for millions of years. The cattle ranches that have replaced some of them require huge inputs of fertilisers, etc, to keep going and there has been an impact, too, on the rainfall and soil erosion in those areas.

Here in Ireland the selective removal of top predator species has had an impact on animal numbers further down the food chain. Huge amounts of deer can survive now that their natural predator the wolf is gone.

Introducing species that have not evolved in harmony with the habitat can cause great upset too. The lush stands of Japanese knotweed can't believe their luck at finding themselves in rich Irish soil instead of having to struggle for existence in the barren volcanic landscape of northern Japan. Mixed deciduous woodland is habitat for many species of birds, mammals, invertebrate and fungi. Monocultures of same-age exotic evergreens are not. Neither are the great stands of rice grass – Spartina – planted on our mudflats in the vain hope that they would reclaim land from the sea. The bare mudflats were good habitat for the former residents of those areas – the flocks of wading birds that probed in the mud for food and could clearly see any perceived danger from all around. They can't feed safely in tall stands of Spartina.

The greater the variety of species a habitat can support, the more stable it is. You don't need chemical analysis of the water to realise that a river with only a few species of invertebrates present has something wrong with it compared to a similar waterbody with the whole gamut of creepy crawlies, from dragonflies and mayflies to water shrimps and leeches. If it is better for wildlife, then it is better for us too.

So, habitat change must be the reason why our corncrake numbers have collapsed over living memory. This is a migratory species that flies here from Africa each summer. Yes, it can fly, despite the fact that it is never seen to do so here as it runs around the wet grasslands it prefers. But flying from Africa is a big effort, and corncrakes are not aerial feeders like swallows who can feed on the wing. They must come down to stock up at intervals on their journey here from sub-Saharan Africa. There was never much feeding on the Sahara Desert, but the corncrake had evolved to cross it in one fell swoop. Now the Sahara is wider because of desertification at its boundaries and many of the corncrakes find it too far to cross in one go and cannot evolve quickly enough to get here by a different route. And I won't even mention the *céad mile fáilte* they receive when they land in a field that is now being managed for silage, not hay with the traditional long grass cover that provided the corncrake with a perfect habitat.

Lots of species and lots of different habitats indicate stability, which is a good thing. But why does the definition of biodiversity extend to the genetic variation within a species. If we have lots of individuals in a habitat, does it really matter if they are all closer than first cousins to each other?

It is this very variation that makes some individuals stronger or weaker or better able to survive change. It is this variation that allow species to adapt to change. It's what allows evolution to happen – the survival of the fittest. If each member of a species is a clone of the next, then when adverse environmental change happens, they, all being the same, would succumb. With genetic variation some individuals are always better equipped to withstand change than others – unless of course that change is catastrophic. The strongest swallows can fly to and from Africa. Those less well able get lost or are too weak to make the journey. It is a case of survival of the fittest. If you can't hack it, you're gone. Those most well adapted to the prevailing circumstances are the ones who survive, breed successfully and pass on their genes.

If there is very little variation, there is not much scope for this. And if there is no variation at all, then an adverse change of environment can wipe out the lot. Hence the worry when the giant pandas in China began to suffer the island effect. These pandas live in bamboo forests. With the growth of human population in China and the increased need for agricultural land, the bamboo forests became 'islands' in vast agricultural areas. Pandas in any one area were confined to their own patch of bamboo forest. Their choice of breeding partner became more and more restricted and inbreeding soon happened. Unhealthy offspring or indeed infertility is a consequence of this, as any cursory study of the Egyptian Pharaonic families will show. Trying to breed pandas in captivity didn't work out either; the pandas didn't take to it. In fact, it has taken in vitro fertilisation to save the pandas from extinction. There are fewer than 2000 individuals left in the wild, which are managed in sixty-seven reserves. A lot of effort and expense has gone into preserving the species.

Scientists have worked out how to remove this genetic variability in species. We can engineer crops so that each plant is a clone of the next. These crops of cotton or soya or rapeseed are designed to be able to grow and thrive in the given set of circumstances they were designed for. Fine if the environment in which they grow is never going to change. Surely this is a short-sighted approach in a world where climate change is so rapidly happening.

The human race is dependent on good biodiversity for its very existence, whether we realise it or not. Yet, we are taking steps to abolish it by making everything the same – abolishing the natural variation that has enabled life on Earth to thrive for the past millions of years. Having less space for wildlife means that some of our ever-increasing population is living much more closely to wild animal species. So it is not surprising that zoonoses – animal diseases and infectious disease-causing particles – can make the jump to humans, causing pandemics.

Increasing biodiversity to make the world a healthier place for all should be a no-brainer.

16

GENETIC MODIFICATION

The spawn of Satan or the best idea ever

I remember when Noel Dempsey was minister for the environment in the 1990s, he was a regular visitor to our wildlife radio programme – *Mooney Goes Wild*. He was introducing the plastic bag tax and it was taking a while because both the grocers and the Department of Finance were not up for such a new tax – although for different reasons. The minister was on regularly to soften up the listeners to be positive to such a progressive tax, which you wouldn't have to pay if you weaned yourself off plastic bags. It happened in the end. Other countries followed our excellent example, and plastic bag-festooned trees, hedges and ditches are becoming a thing of the past. Although in some popular dog-walking areas they have been replaced by 'biodegradable' bags of dog poo, hardly an improvement!

It is now easier than ever for all sorts of points of view to find their way into the public domain, all purporting to be backed by scientific research. In the interest of so-called balance, both sides of an argument must be allowed to be aired even when there is only one genuine, factual side actually to it. A listening public or those following the discussion on social media can have no way of knowing the truth of what is being stated. Get up enough of a head of steam and you can convince people that 5G spreads Covid-19, or that vaccines are terrible and should be avoided, or that spaghetti grows on trees, although the purveyors of that last scam did own up to it being an April Fool's joke. (They had even produced pictures of trees covered with growing spaghetti.) It was the BBC, no less, in 1957, that ran a three-minute news item on the bumper crop of spaghetti that year, thanks to generations of successful breeding that gave rise to such wonderful spaghetti-growing trees. Thousands of viewers were fooled, including the director general of the BBC, and many wrote in to ask where they could buy these trees so that they could grow their own at home.

Along with the fact that spaghetti was a relatively new

food stuff in 1950s Britain and Ireland and that people were not sure if it grew or was made, the magic words that really fooled viewers were 'generations of selective breeding'. This is a term well known to crop breeders, farmers and gardeners alike. Crossing parents with desirable traits could lead to offspring with all those traits in the one individual. And so, by continuing this process, better crops could be produced. Cows with more milk per animal or leaner pigs were all the result of selective breeding of suitable parents.

In Ireland we have always been interested in breeding potatoes that are resistant to disease – particularly potato blight. As a result, all the potatoes in our crop, if we grow our own, or bag of potatoes from the shop will be the same, grown from seed potatoes (small tubers). There are no actual seeds involved. To breed potatoes, which only happens in plant-breeding stations, you need male pollen and the female ovary, all of which are in the potato flowers. Potato plants do have flowers. Those of us that grow them know that the potato crop won't be ready to harvest until the flowers have appeared and then somewhat later the stalks have died off. It is from these flowers that the plant breeders get the pollen to cross with other potato varieties to produce new disease-resistant varieties. The pollen is placed on the female parts of the required cross and it goes on to form potato seeds. These are small apple-like fruit on the stalks, which are gathered and grown on, to see what will be the result of the cross. Of course, the offspring will have a whole range of characteristics – some desired by the plant breeder, some not. The breeder will want the new variety to have whatever resistance is possible to blight, but it will also have to be a good cropper and will have to taste nice and be easy to grow. All these traits in the one new variety take several years to perfect and then enough potato tubers have to be produced for the growers who want a reliable crop. No growing tiny potato seeds for them.

So scientifically what is actually going on here? The characteristics of each potato are carried on the chromosomes present in each cell. A parent potato may be a good cropper or have a pink skin or taste delicious or have a really dry floury tuber – whatever. There will be genes controlling all of these traits at certain places along the chromosomes. Each plant has two copies of each chromosome in every cell, but when pollen is being produced, only one copy of each chromosome goes into each pollen cell. Similarly, the ovary of the other half of the cross will have only one copy of each chromosome. So, it is really chance if all the desirable traits end up on the chromosomes in the pollen that fertilises the perfect ovary and develops into a seed. The cross has to be done many times and the seeds planted out and the new potatoes examined to see if they have the desired traits. It is a long and complicated business getting a potato that has everything.

Genetic engineering is a way of doing this much more efficiently. As genetic studies advanced in the latter half of the twentieth century it was possible to make maps of chromosomes, identify each gene and determine what characteristic in the organism it controlled. The simplest organism is a bacterial cell. It only has one chromosome, and bacterial cells generally reproduce just by splitting so that each new bacterium is a clone of the last. Mistakes can happen in the copying process – mutations – so that some genes vary slightly and this is how change can occur. New varieties of bacteria appear that are mutations of previous ones and can cause us trouble if, for example, they are resistant to common antibiotics, as MRSA is.

But knowing what the gene sequence is in a harmless bacterial species is a great scientific leap forward. Supposing you could somehow put a new gene on to that chromosome, then the bacterium would behave in a different way, and the new cells, all made by cell division, would have the new gene too. Wonderful advances have been made in this way.

Diabetes is a human condition caused by the inability of the body to break down sugar because not enough of the hormone insulin is being produced. This a serious and eventually life-altering condition. Supposing you could make and give someone with this condition insulin to break down the excess sugar. But insulin is specific to each species. Formerly people were injected with insulin taken from cattle and pigs, which did work but obviously could cause allergic reactions. It was taken from each animal's pancreas – the gland where insulin is made. But if you could get human insulin that would be the holy grail.

The gene that controls the production of insulin in humans was identified and, using genetic engineering techniques, it was inserted into a bacterial cell of a harmless variety of E. coli. Lo and behold, cultivating this genetically engineered bacterium led to it producing human insulin. And nowadays people with insulin requirements can inject themselves with human insulin when they detect from a simple blood prick test that they need to. Everyone is delighted about this. So far so good for genetic engineering.

Rennet is an enzyme produced in the stomach of a calf and traditionally was used to form curds in milk as the first stage in making cheese. All cheeses began their lives in this way. The gene for making rennet was identified, inserted on to the chromosome of a bacterial cell and cultivation of the bacteria produced rennet for cheese-making. Eureka. All card-carrying vegetarians could now eat cheese with impunity as no animal had lost its life in the making. All Irish cheese is made using bacterial rennet – look at the ingredients on the wrapping. It is only the really traditional cheeses in really traditional areas that still stick to the calf rennet – such as Italian parmesan or French camembert or Spanish manchego and indeed only some of these cheeses – not all. We all happily eat cheese made with bacterial rennet. Another win for genetic engineering

So why then, if it is such a winner, do lots of people consider it as it were to be the spawn of Satan? Ah well. In the cases outlined above no one actually consumed a genetically modified bacterium. They only got an enzyme or a hormone made by the bacterial cells cultivated under careful laboratory conditions. The modified bacterial cells themselves were not coming anywhere near us. Well then, if that is the case, why was there all the fuss about genetically modified sugar beet that was being trialled by Teagasc in Carlow in the 1990s? Sugar beet is processed to get sugar – a carbohydrate – from its very sweet swollen roots. The sugar is purified and what appears in Siúcre bags is just sugar crystals. No bits of beet or beet cells containing chromosomes – modified or otherwise. If we can eat cheese that has bacterial rennet in the manufacturing process with absolutely no compunction, what is the problem with crystals of sugar?

Ah, but you see, it was not the sugar end product that was causing the difficulties, it was the actual modified sugar beet plants themselves. If you are a farmer growing crops, your harvest depends on how well the crop has grown. One problem facing growers is that their crop in the field will have to compete with whatever else happens there. It will have to compete with the unwanted plants that spring up as well – what farmers call weeds. These can cause a significant reduction in the beet crop as all the plants – wanted and unwanted – compete for water, light, nutrients, space, etc. Farmers clear their field entirely before sowing the crop of beet by spraying with herbicide. But once the crop is established and growing, it could not be sprayed with weed killer anymore, as this would kill the sugar beet too. Until that is genetically modified sugar beet – immune to weed killer – was designed. As the crop grew, the field could be sprayed whenever it was deemed necessary with herbicide, the weeds would die but the sugar beet wouldn't.

This beet had been tested in closed sites and now in the 1990s was being trialled outdoors in Carlow. Many of the vegetables we eat are derived from wild plants and sugar beet is closely related to sea beet, a plant of coastal districts. Pollen in field crops is moved from one plant to another by insect pollinators who visit the flowers seeking nectar. Pollen from genetically modified crops can be carried by insects who visit the flowers that are open at the same time. Modified pollen could be carried to wild sea beet, pollinate it and the offspring would be resistant to weed killer too –

superweeds. Okay, Carlow is far from the sea, but this was a field trial, needed before these modified beet seeds could be given general release. If approved, the crops could be grown anywhere and it would only be a matter of time before wild sea beet was affected.

In the event, the trials were sabotaged by activists who took the law into their own hands. And separately, economic circumstances were causing the closure of beet factories. Tuam had already gone by 1983, followed by Thurles in 1989, but it was EU restructuring that forced the closure of Carlow in 2005 and finally Mallow in 2006. Irish farmers no longer grow sugar beet as a crop for sugar and so that was that.

A different problem with genetically modified crops is the situation with seeds. Traditionally, farmers − particularly in poorer parts of the world − harvested the seeds from their crop and used them to grow the following year's crop. (Hence all the terrible admonishments about the foolishness of eating the seed corn.) Makers of genetically modified crops sell the seed to farmers who are happy to grow them and in whose countries such crops are eaten. However, it is possible to modify such crops in a way that the seeds produced are sterile. To grow next year's crop the farmer has to buy new seeds again. This was considered to be such an injustice by rice farmers in India that the subsequent outcry forced the discontinuation of this terminator gene technology. Genetic modification was not being seen in a good light.

On the other hand, farmers and consumers in the United States and Canada see no problem with eating food made from genetically modified soya, rapeseed and maize. Brazil and Argentina grow large amounts too. Much genetically modified feed is fed to cattle. In Europe genetically modified food for human consumption must be labelled as such, if it contains more that 0.9% approved GM product. There is great consumer resistance to it – particularly in France and Germany. Opponents fear side effects – as yet perhaps unde-tected – but they don't want to be the guinea pigs used for testing, as it were.

Reputable scientists maintain that what is happening is merely a speeding up, in a more precise and focussed way, of what the plant breeders were spending years doing. And, in some cases, they do have a point. Back to potatoes again. Blight is still a major disease of potatoes particularly in Ireland, where warm, muggy summer weather is conducive to its spread and blight warnings are issued with national weather forecasts.

Potatoes are native to the highlands of Peru, where it was a staple food of the Incas and was used by them as far back as 400 BC. The Spanish conquistadores brought the potato to Europe, where it was definitely established in Spain by 1570 and first was traded in the port of Waterford soon after 1586. It grew very well on poor wet Irish soil and its excellence as a foodstuff enabled the population of Ireland to increase from 3 million in 1700 to over 8.5 million by 1845. On 20 August that year, an easterly wind blew spores of a fungal type of organism across the sea from Britain and so arrived potato blight. This disease had originated in Mexico, where it spread from wild potato species to the cultivated crop when it was planted there. It persisted in infected tubers and spread northwards up to Philadelphia by 1843 and from thence to Belgium by 1845 in potatoes that were carrying the infection. It spread from there to England and the easterly wind blew it here. Where were the prevailing southwesterly gales when you needed them?

We know the rest of the sorry story, at least as far as potatoes and famine are concerned. But the organism causing potato blight is still an enormous problem. Because it is a fungal-like organism, it mutates and changes fast. A second strain of potato blight has now come to Europe and joined forces with the original, making it an even faster mutating enemy of potatoes than ever. By the time plant breeders have a new blight resistant variety developed, the blight organism is well on the way to adapting itself for a new attack on it. It is an arms race – with the potato blight holding very strong cards indeed.

Getting disease-resistant potatoes out quicker would reduce the need to spray and spray again potato crops with anti-blight chemicals – surely something we all want. Such potatoes can be produced using genetic engineering techniques, but the very mention of field tests on these is as a red rag to a bull with many people. Are they right to object?

Nothing is ever straightforward. There are two kinds of genetically modified organisms. One kind have had genes taken from another species entirely inserted into them. This procedure is known as transgenic modification. Genes from daffodils inserted into rice gave golden rice – a controversial GM crop. Cisgenic transfer, on the other hand, means that genes are taken from one strain of a species and inserted into another of the same species. There are over 200 wild potatoes species in Peru – many of which are completely resistant to potato blight. Successfully crossing one of these with a cultivated variety would give a blight-resistant potato that would need much less sprays. Blight-resistant potatoes do exist and are grown by organic farmers who do not wish to use sprays, but they are not widely available.

Using traditional plant-breeding methods, crossing a blight-resistant wild potato with a good-yielding, nice tasty potato to get a blight-resistant one, requires five years' work: making the cross, sorting out the progeny, growing on likely ones, building up stocks for growers and releasing them on the market. Meanwhile, the blight organism is changing and adapting too. Genetic studies on potato genes can locate the precise location of the blight-resistant gene, and genetic engineering techniques can insert it into a commercially viable potato in one fell swoop. This is the result the plant breeders want, but they have to go through a trial and error process to get there, which takes much longer.

The cisgenic genetically engineered potatoes have been created and are at the stage of going through field tests before the next stage of licensing for the open market can take place. There is a lot of opposition to these trials. Trying to explain what cisgenic GMOs are as opposed to transgenic takes a great deal longer than the five minutes allocated to that side of the argument in a 'discussion' on radio or television. The so-called discussion is more like a dialogue of the deaf where each side – fully convinced of the right of their argument and the wrong of the other – doesn't pay any attention to what is being said. So how can a listener possibly be educated in the matter with this carry-on?

Mind you, there are other GM potatoes being created in the US, which for example don't display bruising when it happens, because the gene controlling this has been de-activated. So, it's back to reading the label again as the EU is very strict about the labels our food must carry. It is up to the consumer to decide whether to buy them or not. Understanding how the world works requires more and more effort, as knowledge about how it works expands. Aren't you glad you are reading this?

17

WHERE DID WE COME FROM?

Evolution v Creationism

The world began on the evening before 23 October, 4004 BC. This date was calculated from the Old Testament, which was considered to be an accurate historical document, by one James Ussher, Archbishop of Armagh, and published to great acclaim in 1650. This view persisted until the late nineteenth century. What was said in the Old Testament was considered to be the truth and those who challenged it were heretics.

But science was moving apace, and studies in the realm of

geology fixed the age of the Earth at 4 to 6 billion years old – a far cry from the 5654 years of age that Ussher postulated. In fact, there was an explosion of scientific knowledge in the second half of the nineteenth century and three men, among others, are responsible for demonstrating how the world works. They were Charles Darwin (1809–1882), Gregor Mendel (1822–1884) and Alfred Wegener (1880–1930).

Charles Darwin outlined his theory of evolution in his book *On the Origin of Species* in 1859. Boy did he unleash a storm of controversy! He declared that all living things on Earth had evolved from earlier forms by a mechanism which he described as natural selection. Over generations of any species, there would always be individuals that were better adapted to life and so they were able to leave more offspring with those same traits. These too would improve in subsequent generations, particularly as the surroundings in which they lived changed – the survival of the fittest in other words. Those that couldn't adapt did not do well, left fewer descendants and were outcompeted by fitter individuals. There were no nursing homes for slow cheetahs or for short-sighted eagles. The fastest and the keenest of eye got all the food and were able to attract mates and breed successfully. Nature didn't tolerate eejits, although of course he didn't put it quite like that in his scientific book.

It went down like a lead balloon among many of the

notables of the day. They knew that God created every living thing and indeed created man on the sixth day and put him in charge of his creation. Had God made imperfect things that needed to improve? And man – man was made in the likeness of God in the Garden of Eden. (It was the woman's fault that he was driven out!) How could Darwin possibly have the temerity to suggest that man too had evolved from earlier hominid species that were now extinct, but that had the same lineage as apes and monkeys? No, it was just too much. And, indeed, to this very day, creationism – the theory that all species on Earth, and particularly man, were created by God as they are now – is still sometimes taught as 'science' in some parts of the United States and in some Islamic countries as well.

The discovery of fossils that demonstrate earlier forms of life and the use of this knowledge as evidence supporting evolutionary theory was dismissed out of hand. God had created all those fossils too at the same time as he created the Earth and put them in the rocks. Surely the scientists weren't saying that God had made a mistake and did away with dinosaurs. No, the whole lot were always there – created by God as fossils. Although there were some dissenting voices wondering why God had done this, why he had tried to deceive humans with fossils that might give rise to the erroneous theory of evolution. Surely God was not a deceiver? It was all very complicated and bewildering.

But the scientific discoveries continued apace. Alfred Wegener put forward the theory of continental drift. He maintained that the position of continents as we see them today is not the position they were always in; that they actually move at the same speed as your fingernails grow. South America was once joined to Africa for example. And even earlier still, 300 million years ago, all the continents were joined together in one super continent which scientists called Pangea. Pangea lasted for 100 million years before it began to break up. There was lots of life on Earth at that stage. Even mammals had begun to appear, although they were only the more primitive marsupials that did most of their growth and development after birth in their mothers' pouches.

The great Pangea landmass first of all split into two landmasses, and the southern one – Gondwanaland – drifted away, carrying its cargo of plants and animals. Australasia, Antarctica and South America were the three continents in this land mass and they soon separated from each other. Antarctica over the South Pole was so cold that no land mammals there survived. Australasia – mainly Australia and New Zealand – never came into contact with the northern hemisphere again and so its mammals remained as marsupials and became adapted to the conditions there. There were never any placental mammals there (except for a few bat species that flew there) until humans brought them in the eighteenth century with catastrophic consequences for the native marsupials.

Evolution of mammals continued on the other section of Pangea, now named Laurasia, which consisted of North America, Europe, Asia and Africa. It had started originally in Africa but spread from there, and these placental mammals – which were born fully developed rather than having to spend time developing in their mother's pouches – soon outcompeted whatever marsupial species existed in the other parts of Laurasia. South America was still wandering about with its cargo of mammals, both marsupials and indeed placentals which had arisen there in the intervening millions of years. Eventually – a mere 3 million years ago – it joined up with North America and its mammal populations were able to travel to and receive visitors from there. This was not good for the poor marsupials of South America – the visiting placentals took over the place, although there are still some marsupials such as the Virginia possums, which did manage to move north and take up residence in North America. There are also many species of opossums in South America to this day, but the other mammal groups are all placental mammals.

One group of mammals – the great apes – never made it to the rainforest of South America. There are monkey species there but none of the great apes (species with no tails). Orangutans are native to the rainforests of South-east Asia and the gorillas and chimpanzees are African natives. When it became apparent that Darwin's theory of evolution included humans – that we too were but mere animals evolved from earlier ape species that had also given rise to chimpanzees and gorillas – well, that was the limit. Humans were specially created and given an immortal soul and could look forward to a life in the hereafter. Animals had no souls and were gone when they died. No, it was just too much altogether.

Gregor Mendel's work was the first to demonstrate a mechanism by which such a thing could happen. He did experiments in his monastery garden in Austria and showed by growing peas and saving their seeds and replanting them, that characteristics in peas could be inherited and passed on intact several generations later. He called his studies genetics. He published his results in 1865, but it didn't attract scientific interest at that time. He became Abbot of his monastery in 1867, and other matters took up his time and attention from then on. He died in 1884, without ever getting the honour he deserved for his seminal work. It was to be 1900 before several independent researchers in different European countries encountered Mendel's work which showed that he had come to the conclusions that they were only coming to now, almost forty years previously.

It took most of the twentieth century to uncover the complete picture. Every living cell has a nucleus, in which there is a set number of chromosomes for each species. The genetic information that determines the characteristics of that species is carried on the chromosomes. The place where each piece of genetic information is carried is called a gene and chromosomes have lots of genes. Humans for example have 46 chromosomes on which there are tens of thousands of genes. The 46 chromosomes are arranged in 23 pairs. We get one of each pair from each parent – 23 from our mother and 23 matching ones from our father. They pair up at fertilisation, and every cell in our body contains these 23 pairs with the exception of our reproductive cells. Egg cells and sperm cells only contain 23 chromosomes – one from each of the 23 pairs. The thing is that the chromosomes in the egg cell are not the same 23 the woman got from her mother, but a mixture of maternal and paternal material and it is the same in the sperm cells which only contain 23 chromosomes too. The offspring produced will have a different genetic complement to either its mother or father or indeed its grandparents. It will be a mixture of all these that goes to make up this individual. And it is the same for all living things that are created by sexual reproduction (as opposed to asexual reproduction, which is merely cell division).

So this is the mechanism by which offspring can differ

from their parents. There are no two people with exactly the same genetic make-up, except identical twins who are clones of each other. This is the same in all other living species too. All individuals slightly differ from each other and they react in slightly different ways to the environment in which they live. If this environment changes, some will cope better than others and thrive. Over the millions of years life has been on this planet, living things have changed and adapted to survive – what Darwin called evolution. When surroundings change slowly, evolution can keep up with it and species survive. If the environment changes too quickly then extinction beckons.

The Natural History Museum of Ireland on Merrion Street in Dublin boasts wonderful skeletons of a huge deer species – the males of which have enormous antlers. So many of these skeletons were dug out of Irish bogs that the species – which had a European distribution – is called the Giant Irish deer. It roamed the tundra regions at the end of the last Ice Age, feeding on the grasses that were able to grow as the ice retreated. It was their heyday in Ireland from 11,750 to 10,600 years ago. They must have been a magnificent sight roaming the country in great herds. They inhabited a very limited world. Their movement northwards was not possible because it was too cold there and the plants that grew in those conditions could not provide enough nourishment for them. Their movement southwards was impossible too because in those slightly warmer areas great forests grew and their enormous antlers impeded movement there. They had 750 years of glorious life here and then the rich grasslands were covered with woodland as the Ice Age came to an end. Huge antlers got you the most wives and ensured that you left lots of offspring with equally enormous female-attracting antlers. No good in the end, though, if you ended up caught by those same antlers in the woodlands that were becoming ever more abundant. The smaller, neater Red deer could manage and still can in the Ireland of today while the Great Irish deer is now just a set of impressive skeletons in natural history museums.

If environmental change happens at a slower rate, then survival of the fittest works. For long periods of the planet's history the climate was such that vast belts of rainforests straddled the equator all around the globe – the entire width of South America, the entire width of Africa and all across the equatorial regions of Asia. These forests were home to great apes in Africa and Asia who lived entirely in the forests, finding all their requirements there. But about 5 million years ago the ever-changing world climate became too dry for the forests of Eastern Africa to survive. Great savannah-like plains replaced them. Vast herds of grass-eating animals roamed the plains providing food in turn for the fleet-footed carnivores – lions, cheetahs and the like. This was not a land with enough fruit and berries to maintain the apelike creatures that had lived so well in the forests. New skills were needed to be able to survive. Standing upright – rather than moving on all fours – enabled greater distances to be surveyed and enabled the front pair of limbs to be used entirely for food gathering, self-defence and tool-making. Greater speeds are achieved on four legs than on two, so catching prey involved working in groups and being able to communicate with each other to do so successfully. Raw meat is very difficult to digest, so the discovery that cooking it makes it more digestible and being able to make fires when required for cookery was a great leap forward. One species of ape-man succeeded the next from the earliest hominid *Australopithecus* – known as Lucy, whose skull found in Ethiopia was 3.2 million years old – through *Homo habilis*, then *Homo erectus* who moved out of Africa eastwards to Asia. *Homo neanderthalis* moved to Europe 230,000 years ago and was contemporaneous and made extinct by *Homo sapiens* who was on the scene from 195,000 years ago. We are the only *Homo* species now in existence.

Discovering all these fossils and piecing the story of evolution together was very exciting work that took place over the second half of the twentieth century and continues today. The better our scientific techniques, the more we can learn about where we came from. We have certainly been around a lot longer than the six thousand-odd years of Archbishop Ussher's calculations.

But how have we been behaving as a species? All the other species on Earth lived in harmony with their environment. They either evolved slowly as the environment around them changed slowly or else became extinct if environmental change happened quicker that they could keep up with.

Why should *Homo sapiens* be any different?

Actually, we are not that different. The rules of ecology apply to us too, regardless of how we try to wriggle out of them or pretend to ourselves that we are wonderfully aware of our impact on the world and know how to tread lightly on it.

18

RECYCLING

Or just an excuse for woeful waste?

When people are asked what they are doing to save the planet – those of them, that is, that feel they should be doing something because the planet needs saving – in a great deal of cases they say that they are recycling. So that's all right, then? Well, it is not. What is all this stuff that they have that they feel will not be a misuse of the Earth's resources if they recycle it? It's grand to buy a plastic bottle of water as long as you put the empty bottle in a recycling bin. I fear the cart is before the horse.

It was in 1994 that a photo was published on the front

page of the *Irish Times* showing a curlew with a plastic six-pack can carrier entwined around its beak and head. The photograph highlighted the scourge that these discarded plastic can-carriers presented to bird life and wildlife in general. Once entwined around head and mouthparts, the creature so adorned would die a lingering death from starvation. The Irish company that made these can-carriers was shamed into action. The Teachers' Centre in Blackrock was approached by them and asked to encourage schoolchildren through their association with Irish teachers to collect the plastic can-carriers, which could be recycled by the factory and made into new ones. The approach resulted in the Ringo Project. This was an educational project, which I drew up for fifth-class pupils in primary schools and provided a series of lessons about plastic and packaging as part of the Social, Environmental and Scientific Studies curriculum which was about to be introduced into schools. For twelve years from 1995 until 2006 this Ringo Project educated fifth-class pupils and indeed the entire school where the project was carried out, on plastic and recycling. Many bags of discarded ring carriers were collected and returned to the factory, where they were reprocessed into new ring carriers. Pupils learned about the whole subject of waste in Ireland and about the three 'R's that were in vogue at the time – Reduce, Re-use and Recycle. The project went online in 2007 and I no longer visited individual schools, exhorting them to collect and recycle ringos, as the can-carriers were called. And then, as can be the case when something goes online and is available free to all, the project sank without trace.

But it had run for twelve years, and it was reckoned in the summation of the effects of the scheme in 2007 that 250,000 primary-school children were in schools during the period the Ringo Project was carried out. The eldest of this cohort – who were in fifth class in 1995 – are now around thirty-six and they range back in age to twenty-four years of age; millennials, I believe they are called. The very group that drink bottled water from plastic bottles and have takeaway coffees in disposable cups, etc. They may or may not place the containers for recycling, if some of the containers are indeed recyclable. The EPA reported that in 2018 – the most recent available figures – over one million tonnes of packaging waste was left out for collection. What are we like? There are fewer than 5 million people in Ireland, and packaging – plastic, paper, card, etc – is quite light, so this is really a huge amount of waste.

Back to the three 'R's. In fact, there should be four, or even five, and the first one should be REFUSE. Refuse to have anything to do with useless material that has to be discarded immediately. If nobody took it, then it wouldn't be presented to us. Next in the hierarchy is REDUCE. Which means taking less of it. I grew up in the 1950s, when plastic hadn't been invented for general packaging use and we all managed to live grand. Plastic is made from oil – the same oil that is in oil wells – and it is chemical processes that make it into plastic. So, a fossil fuel is being used to create a material that cannot be broken down or recycled as in the case of many soft plastics and the commonly used polystyrene. Refusing and reducing drastically our use of these plastics in particular has to happen urgently. Plastic-wrapped fruit and vegetables suit the supermarkets. We don't *have* to buy them.

REUSE is the next R in the waste hierarchy. In wealthy Western Europe and North America, we don't go in much for this, nor are we encouraged to do so. How can manufacturing industries thrive if people get many uses out of their stuff instead of quickly getting bored of it and wanting new? Some of the statistics on how often some people wear fashion items are truly frightening. Many fabrics in the polyester field contain fibres made from oil, so more use of finite fossil fuels. Getting longer out of our possessions – considerably longer – puts much less strain on a finite world. We should buy fewer (albeit sometimes more expensive) things that are made nearby rather than cheap items that have to travel many airmiles.

And then there is RECYCLING. What this actually means is that the item, if it is suitable, goes back to be a raw material for a new manufacturing process. Much energy is needed for this. Plastic water bottles, when cleaned and separated from their caps and any labels, can be shredded and made into fibre for use in fleeces, hoodies and garments like that. Try ironing one if you don't believe me – it will melt and stick to the iron if it's any way hot at all. Before the 1980s we all managed fine without needing to buy and carry around plastic bottles of water and nobody died of drought. It wasn't the Sahara here then, nor is it now – not yet anyway. Water is expensively cleaned and made available for us to drink. If we must carry it around, simply fill a permanent bottle from the mains supply. Imagine, 183 million litres of bottled water were purchased in 2017 by fewer than 5 million of us, many of whom refuse to pay for the luxury of the clean drinking water that is piped into our homes. Ah but some of us recycled some of the bottles, so that's all right then.

Composting is another word for recycling. Organic or food waste is broken down by organisms such as fungi, bacteria and worms and thus enriches the soil so that new plants can grow. We can do this ourselves in compost bins in the garden or use the brown waste facility provided by local authorities. In the presence of oxygen, these organisms break down the waste and carbon dioxide is given off, but new growing plants re-absorb it again in normal circumstances. If it ends up in landfill, however, that is entirely a different story, which it will if it is put in the black bin.

Another R is RECOVER. This means get what can be got from what would otherwise be wasted. Plastic is made from oil, which burns very well Instead of being dumped, left as litter or thrown carelessly into black bins that go to landfill, the bottles could be burned carefully at high temperatures and the heat used to make electricity by heating water for steam. Ah but we couldn't be doing that now, could we? That would involve using incinerators and we don't like those in Ireland. We do have a few in Ireland, built to very high European standards and monitored to make sure they operate properly, but some people just can't be convinced of their worth. They would rather their plastic bottle went anywhere else – even to a landfill.

What is the story on landfills – or dumps or super dumps, as they used to be known? This is the worst way of getting rid of stuff we don't want. Disposal of it in a hole in the ground, compressed to remove all air and then covered and left. This is still the fate of the contents of our black bins – known as household or municipal waste. In 2017, the latest year for which records are available, we landfilled 23% of our waste – some 637,000 tonnes. Why are we putting this amount of stuff into our black bins and paying through the nose for this most unsustainable practice – or indeed not paying for it but dumping said waste around the countryside. Why do we have so much of it? If we have been really careless and there is food waste – organic waste – in these black bins, the helpful decomposing organisms, which work so well in the garden compost bins, cannot operate. They need oxygen to do so. There is no oxygen in squashed-down landfill sites. Instead, there are other bacteria who can work in the absence of oxygen and these break the organic waste down, producing methane. This is twenty-four times more efficient at causing global warming than carbon dioxide is. Unmanaged dumps emit this into the atmosphere and contribute to climate change. It wouldn't kill us to keep all this waste from the black bins, but it will if we don't.

So, as I explained endlessly for twelve years to very interested fifth-class pupils, recycling is not the solution to the huge amount of waste produced by ourselves. Producing much less of this type of material is the answer, resulting in us having less stuff, having better stuff made locally and keeping stuff longer. A whole generation were taught this in primary school (along with Irish and hard sums).

Learning about the environment in school is necessary, but keeping up the learning and acquiring knowledge and understanding all through our lives is essential. Ten years is all we have left if we continue as we are going. They tell us we can still change this trajectory, but only if we know how vital it is that we do it and insist on all steps being taken to achieve it.

19

AND THEY ALL LIVED HAPPILY EVER AFTER

The End?

Yes, well, maybe we should contemplate the end. Whatever did happen to that species *Homo sapiens?*

It was a species like any other on planet Earth for most of its existence. It evolved very fast in response to the Earth entering a phase where it warmed and cooled in a whole series of ice ages two to three million years ago. The forests it lived in in East Africa became savannah through lack of moisture. The hominid species that led to the evolution of *Homo sapiens* evolved to be able to live here. Omnivory meant that all available food could be digested so they weren't beholden to any one food group. Standing erect, rather than using all four limbs to move, meant that they could see further over the plains and had two limbs free to carry things and use tools.

Capturing faster four-legged animals for food meant that they had to co-operate and communicate with each other, so acquiring the ability to speak was a great leap forward. Controlling fire so that food could be cooked meant that it was much more digestible, so *Homo sapiens* grew big brains. This was a complicated way of life, so they had to live in groups, and teaching their offspring all of this took many years – much longer than it took for the young of other species to earn all the tricks of the trade.

But it worked. Slowly, agonisingly slowly, their numbers increased, although looking at the genetic variation among them all it was a close run thing. They seem to have been down to one female at one stage, mitochrondrial Eve, as this is the only way to account for how genetically similar they were. They weren't identical of course and they adapted to live in the different parts of the world to which they spread from Africa – Australia, Asia and Europe.

But they were still controlled by the rules of ecology. They were all hunter-gatherers and moved about to ensure they had enough to eat. It was a hard life and their expected lifespan wasn't much more than thirty years. They had a slow reproductive rate as their infants grew slowly, so it would be three years at least between births and that would further depend on there being enough available food for a considerable period of time for the females to become pregnant, carry the child for quite a long gestational period and feed it with her own milk for two years or so afterwards.

But they managed. They had some good times, too, if their cave drawings of the high spots of their hunts are anything to go by. And by 10,000 years ago they were in every continent except Antarctica and there were probably about 10 million or so of them around. And then they created farming. They discovered how to grow their own crops and domesticate animals for meat. Their supply of food was much more assured, so many more of them survived. Ten thousand years ago they were the same sort of people as in the twenty-first century – just as smart, and just as well able to solve problems and come up with solutions. Farming with stone tools and burning off unwanted vegetation was all very well, but in many continents – although not in Australia – they discovered metals that made much more effective implements to till the soil. They could clear more and more land for their crops, which they needed to do as their numbers were increasing all the time. In fact by the twenty-first century they had chopped down and removed half of all the trees that existed in the world before farming happened.

This of course meant that there were fewer places, much fewer, for all the other species with which they shared the planet to live. But they seemed to be quite oblivious to this. The metals could be fashioned into weapons to make killing and butchering animals more effective. They could also use them to advance their own situation vis-à-vis their fellow man. They learnt how to use the natural resources of their world as energy. Harnessing the wind meant they could sail greater distances than rowing could ever achieve. Timber as charcoal enabled them to smelt metal, and the discovery of coal improved and speeded up this whole process. Darkness was still a problem when the sun went down, but steam engines in ships allowed them to hunt the great whales for their oil to burn for light – a task they took to with gusto.

And they broke the first rule of ecology and hunted these great creatures almost to extinction. No survival of the fittest here in such an unequal competition. In fact, it was only the discovery of mineral oil in the ground in the 1800s that saved whales from extinction. Discovering this type of oil was a game-changer, and together with great resources of coal and, subsequently, natural gas, humans were able to utterly transform the world. The fact that burning all these fossil fuels released large quantities of carbon dioxide into the atmosphere wasn't apparent to them at first. When they did find out about it, many people refused to believe it, notwithstanding the fact that all that extra carbon dioxide in the atmosphere (it had gone from 270 parts per million to 417 ppm – an increase of over 50% in less than two hundred years) was causing the world to warm up at an alarming rate.

They over-hunted many other animal species too. They got very good at catching fish in the oceans, using huge ships and finding shoals using sonar, and soon these fish species became exceedingly scarce. They rode alongside their trains across North America, shooting the herds of buffalo that roamed there – just for the craic. They didn't need or use them as food.

But it was when they learned how to stop the natural brakes on their numbers that their impact on their planet became really serious. While farming had greatly increased their supply of food, they still suffered from many diseases. Infant mortality was high and plagues killed great numbers of them. The Black Death in the fourteenth century killed half the population of Europe at the time. The so-called Spanish flu killed millions more in the early twentieth century. But science discovered the organisms that were causing these catastrophes and vaccines were created to deal with these diseases and others like them. Antibiotics and anti-viral medicines could treat those misfortunate enough to have illnesses and injuries where bacteria and viruses were life-threatening. And this all worked.

The population of *Homo sapiens* reached 1 billion in 1804, and by 2020 it had increased to 7.8 billion. Gone up by a factor of over 7 in just 216 years. Poor planet Earth was still the same size, so to have space for all these humans, other species had to give way. In 1970 there were 50% more individual wild animals on Earth than there were in 2020. That was some clear out of fellow species.

Many humans seemed to think they needed lots of possessions in order to live, but they created huge amounts of waste in acquiring and indeed relinquishing these same possessions when they got tired of them. Much of this waste ended up in the oceans and didn't break down or decompose because none of the normal decomposers – bacteria and fungi and the like – that had been breaking down and recycling the nutrients of organic waste could cope with the new-fangled waste such as glass, aluminium and even more insidiously plastic. Much of the oceans became intolerable for fish to live in. Water reserves were polluted with waste disposed of on land. Enormous use of and pollution of fresh water supplies made life unsustainable for those living in regions where the climate change caused by global warming increased drought conditions.

And what happened in the end?

A new pandemic appeared in 2020 for which initially they had no treatment, cure or way of preventing, but they were able to pull together to save their skins. Everyone saw how important it was to avoid getting this disease, and decisions taken by governments at country levels were obeyed. The world population did eventually rally.

And did they not take action to prevent climate change destroying their world and making living conditions impossible? Ah, that was much harder. Many who should have taken decisions at government and country level refused to see the enormity of the situation because of their wealth. They seem to have been unable or unwilling to bring the people with them for the good of long-term goals and ultimately humankind itself. The disruption such decisions would cause to the economy, as they called it, was too high a price they felt. And did ordinary people who were being affected not make them do it? Well, of course climate change was not happening as noticeably in the countries where these rich people who were emitting most of the carbon dioxide and creating all the waste lived. It was in other countries, poorer counties further away, where the greatest effects were happening. And their governments hadn't great clout on the world's stage when decisions to reduce human impact were being taken.

So?

So that's it. We know about it; we know what is causing it; we know what will fix it; we think we cannot afford it – our economic situation won't allow it. Ten more years of living as we do and the decision will be made for us. What is one species more or less to planet Earth?

SAVED IN THE NICK OF TIME

… Or maybe not. Have the human race copped on just in the nick of time? They have been done no favours by the forecasts that always said there was still time left before the point of no return was reached. This was going to happen by the end of the twenty-first century, then it was going to be by 2050. Finally, in 2020, it was definitely stated that by 2030 the tipping point would be reached if there wasn't a great reduction in the amount of carbon dioxide being emitted. Up until that declaration there was always still time to do something in the future, but never now. In the words of the famous prayer by St Augustine, 'Lord make me pure – but not yet.'

The Covid pandemic demonstrated just what an impact so-called normal life had been having on global emissions. The first lockdown – in the first half of 2020 – led to a global reduction of 8.8% in carbon dioxide emissions in the areas of ground transportation, power generation, industry and aviation. The annual amount of carbon dioxide emitted into the atmosphere had more than doubled from 1971 to 2016, and 91% of this came from burning fossil fuels. (The other 9% came from changes in land use.)

The carbon dioxide emitted in these four areas can be reduced by human effort, and it is essential that this happens if we want to survive as a species. Some countries in Europe are already leading the way in the area of power generation. According to Eurostat figures for 2019, all of Albania's electricity is generated from renewable hydro energy. Ninety-eight per cent of all electricity in Norway is generated from renewables – mainly hydro, with a small amount from wind. Denmark leads the way in wind generation, with almost 60% of their electricity generated in this way. The world's largest offshore wind farm is currently being built 131 kilometres off the northeast coast of England on the Dogger Bank. It will generate enough electricity for 4.5 million homes in Britain. In

Ireland we get 35% of our energy from renewables, mainly wind with some hydro. The rest is from burning fossil fuels – including incredibly, until just recently, our own indigenous fuel peat. We have the greatest amount of offshore wind in Europe, both off the Irish Sea coast and our Atlantic coast. We could, and should, be leading the way in harnessing this for electricity generation and indeed for export to the rest of Europe. Given that 17% of our greenhouse gas emissions still comes from this sector, notwithstanding the fact that we have more than a third from renewables already, it could make a significant impact, not only on our own balance sheet but on that of other European countries too.

Therefore, it is interesting to see the reaction to a proposed scheme to build a wind farm offshore on the Kish and Bray sand banks, ten kilometres off the coast of south Dublin and Wicklow, where over one million people live. The proposals state that it will be capable of generating enough clean renewable energy from the wind for over 600,000 homes and therefore decrease the dependency on electricity generated by burning fossil fuels. Great emphasis in the reaction to this is placed on the change in views out to sea from houses along the coast – ignoring the fact that if reductions are not made in greenhouse gas emissions, rising sea levels will have put paid to these same dwellings by the end of this century. Taking the EU as a whole (EU 27) in 2019, 43% of electricity generation came from burning fossil fuels. The rest of the generation came from renewables and nuclear. We need to increase our figure of 35% pronto.

Mind you, it is a fact that generating electricity from nuclear energy results in no greenhouse gas emissions whatsoever, and we have enough uranium for 200 years more of electricity generation. Britain is currently building an exceedingly large nuclear power station, so they view it as a way to end generation of electricity by burning coal. It is seen as a medium-term solution, to allow us to make up the difference between what is generated by sustainables and what otherwise is produced by burning fossil fuels. In the long term there could be a transition to nuclear fusion, which is non-radioactive. Seventy per cent of France's electricity is nuclear generated, while 35% of Belgium's also comes from nuclear.

Ground transportation was the area in the first half of 2020 where the greatest reduction in the emissions of carbon dioxide occurred. In Ireland a full 20% of our greenhouse gas emissions come from this sector. However, as only essential travel was permitted during the first Covid-19 lockdown in 2020, there was a great shift to working from home with a resultant large drop in road traffic. So it can be done. Add to this a change in the type of vehicle, from fossil-fuel burning to electric battery-powered, charged with electricity generated from renewables, and the way to reducing emissions is very clear. But as long as senior ministers fail to set an example by having electric cars themselves, citing the lack of a countrywide charging network, change and improvements will be slow in coming. This lack of charging points should be addressed pronto. Public car parks and supermarket parking areas in Britain all have charging points installed, which makes it more manageable to have an electric car there. It we were really serious about slowing carbon emissions from the transport section, public transport would be ramped up so that car dependency could drop substantially.

In 2020 there was a surge in the number of cyclists, a clear indication that in the summer months at least, people were willing to leave the car in favour of two wheels. But it must be perceived as an absolutely safe way to travel, both for children as well as adults and it definitely must not impact on pedestrians by forcing them to share the same route as speeding silent cyclists. There has to be a policy of providing safe cycling routes that do not compete with motorised traffic and this will mean reducing the amount of traffic lanes in many of our cities and towns. Decent public transport, lots of safe cycling lanes, electric cars – it can all be done.

The sector in Ireland that is the biggest emitter of greenhouse gases is the agricultural sector. A full third of all our emissions come from this sector. This is because we have such a large cattle industry in particular. Greenhouse gases come from the cattle themselves in the form of methane, as well as from their slurries. And in parts of the world where forests are cleared to create ranches for cattle, or to grow soya to feed them, the impact is even greater. While humans are naturally omnivores, we do not need anything remotely like the amount of meat produced globally to sustain us. In order to stop runaway climate change, the way we produce our food and what we eat has to change to a way that is dominated by plants. As long as there is huge demand for meat then it will continue to be produced, so reducing the demand is the way to slow it down. And research continues into ways of growing crops not using soil at all, which means more food can be produced.

Removing carbon dioxide from the air by photosynthesis – in other words by growing trees – seems to be a no brainer. Yet tree planting in Ireland continues to be stymied by restrictions and red tape. State forests are managed by Coillte – which is a commercial semi-state body. The Government, i.e. the Department of Forestry, does not plant or fell trees, contrary to what some people here believe. They issue licences for tree felling and planting and oversee the awarding of grants and the conditions attached to these. While we had an extremely low percentage of the country covered in forests and woodlands a century ago, a mere 1.2% in 1928, afforestation – which is planting trees in areas which were not forested – has meant that 11% of the country is now under trees. (And it is estimated that hedgerows and small groups of trees cover another five per cent.) Much of this afforestation was carried out by Coillte, who up to 1995, were afforesting on average about 15,000 hectares a year, thus annually increasing the amount of forest cover in Ireland.

However, in 1995, changes in the rules governing state aid meant that Coillte could no longer get grants from the Department of Forestry to plant trees. So, they were no longer able to afford to purchase land for afforestation and no new forests have been planted by Coillte since this time. They do, of course, manage the existing forests that they own, harvesting the trees when they are mature and replanting. This is called reforestation. They harvest trees that may have been planted forty or fifty years previously when planting regulations were entirely different. Nine thousand hectares of forest are harvested by Coillte each year. If some of those forests were planted in less enlightened times on blanket bogs, then Coillte does not replant these areas. There has been no planting of trees on blanket bog by Coillte or anyone else for the last forty years. If the land is suitable then replanting takes place, but there are very specific guidelines for what can be planted and in what amount. Planting licences specify that even where conifers are planted, a proportion of the site must be planted with broadleaved trees. Trees must be a certain specific distance from streams and waterways. Some of the site must be left unplanted to allow for biodiversity of other plant and animal species.

So, if Coillte no longer carries out afforestation, is it happening at all? Any new land that is planted with trees is now done under private ownership – much of it by farmers, who also are bound by current tree-planting guidelines. The rate of afforestation is greatly slowed down to not much more than 1000 hectares in the 2019 season. This rate has to increase to make any impact at all, as it takes about fifteen years for a newly planted area to start sequestering carbon.

Can we improve our biodiversity? We can of course if we think that it is important. A great example of what can be done is the restoration of the mountain gorillas in the Virunga Mountains in Rwanda. Here the forests where they lived were under pressure from farmers needing more land to grow crops and the animals themselves were falling prey to poachers. Forty years ago, there were only 250 animals left and the species was on the verge of extinction. But enlightened policies were rolled out from 2005 onwards that benefitted the local people as well as the gorillas. Tourism was strictly controlled, and rangers were employed to track and protect every gorilla group. The money paid by the tourists for a permit to be escorted to the presence of gorillas just for one hour (500 dollars a head when I visited there in 2009) goes to the communities in the area. It is in their interests to make sure that the gorillas thrive, and there are now over 1000 gorillas in Virunga Mountains today.

These gorillas pay their way, so it is worth it to conserve them. Are we getting to the stage where we know the price of everything, but the value of nothing? Describing our environment in terms of the ecoservices it provides certainly indicates this mindset. Our pollinators are worth X amount of billions of dollars per year. Boglands work as precious sponges to stop our valuable homes being flooded. If the end result is that of biodiversity being conserved and improved, should the means be questioned? What is in it for us? Is that what we have come to? There is no other species available that can or indeed would pay good money to conserve the human race, so it is up to us to do it for ourselves, while it is still possible, so that we can ultimately justify our moniker *Homo sapiens*.

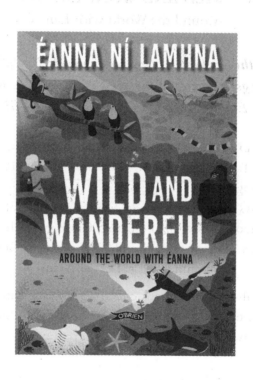

OTHER BOOKS BY ÉANNA NÍ LAMHNA

WILD AND WONDERFUL
Around the World with Éanna

Glow-in-the-dark owls, eggs boiling in Icelandic hot pools, the gangster tactics of the devil's coach-horse beetle – Éanna Ní Lamhna has seen them all!

Éanna explores the wonders of our wild world, from a safari in Tanzania to the cloud forests of Costa Rica, from rat-hunting in Canada to whale watching in New Zealand. She draws on her experience as a diver to tell of face-to-face encounters with fascinating fan worms, elusive sea hares and a murderous crab, and rings the alarm bells on the environmental challenges facing us.

Éanna also recounts with cheerful relish the pitfalls and delights of being a broadcaster and a scientist. Sure why would anyone want to be anything else?

'Charming ... explores the weird and wonderful sides of Planet Earth.'

Irish Independent

WILD DUBLIN

Exploring Nature in the City

"As readable and attractive a nature publication as I have seen..."
Village

Éanna Ní Lamhna

Photographs by Anthony Woods

WILD DUBLIN

*Above buildings, beneath rivers and canals, amidst
bushes and trees ...
inside the M50 nature abounds*

Minks in the Dodder, whales on the coastline, bats in Raheny,
newts in Dundrum, badgers in Rathfarnham, otters in
Ringsend – these are just some of the fantastic creatures to be
seen in the capital.

Éanna's intriguing running commentary both entertains
and explains, and this is a book for the entire family with
exclusive photographs by Anthony Woods and specially
commissioned watercolours by David Daly.

*A truly stunning book, full of fascinating facts and
beautiful photos, showing a city we thought we knew in a
new light.*

'Éanna Ní Lamhna has provided a vivid account of the urban
flora and fauna and covers a prodigious range of subjects.
Words and pictures fuse seamlessly in this lively celebration of
nature in Dublin.'

Irish Examiner

ALSO FROM THE O'BRIEN PRESS

THE GREAT BIG BOOK OF IRISH WILDLIFE
Juanita Browne
Illustrated by Barry Reynolds

A beautifully illustrated book that guides the reader through the seasons in Ireland

Explore nature in your back garden as well as in mountains, rivers, forests and sea. Learn about weird and wonderful natural phenomena, such as the metamorphosis from tadpole to frog; the red deer rut in autumn; or a starling murmuration in winter.

The perfect introduction to Irish birds, mammals, plants, insects, and amphibians, overflowing with stunning photographs and engaging, child-friendly cartoons.

'This book reads like a diary of the seasons … in the garden, local park, forest or beach. There are practical tips for helping animals make the most of their habitats, as well as reminders of what animals and environments are best left alone. Barry Reynolds's illustrations are intermixed with photographs that reinforce the amazing wildlife all around us, even in urban environments, and the role that children can play in ensuring they thrive.'

Irish Independent